Lecture Notes in Information Systems and Organisation

Volume 21

More information about this series at http://www.springer.com/series/11237

René Riedl · Fred D. Davis
Rajiv D. Banker · Peter H. Kenning

Neuroscience in Information Systems Research

Applying Knowledge of Brain Functionality Without Neuroscience Tools

 Springer

René Riedl
University of Applied Sciences
    Upper Austria
Steyr, Oberösterreich
Austria

and

Johannes Kepler University Linz
Linz
Austria

Fred D. Davis
Information Systems
Texas Tech University
Lubbock, TX
USA

Rajiv D. Banker
Fox School of Business and Management
Temple University
Philadelphia, PA
USA

Peter H. Kenning
University of Duesseldorf
Düsseldorf
Germany

ISSN 2195-4968   ISSN 2195-4976   (electronic)
Lecture Notes in Information Systems and Organisation
ISBN 978-3-319-48754-0   ISBN 978-3-319-48755-7   (eBook)
DOI 10.1007/978-3-319-48755-7

Library of Congress Control Number: 2016955914

Printed on acid-free paper

This Springer imprint is published by Springer Nature
The registered company is Springer International Publishing AG
The registered company address is: Gewerbestrasse 11, 6330 Cham, Switzerland

Contents

Part I
Applying Knowledge of Brain Functionality Without Neuroscience Tools: The Approach

Chapter 1
Introduction

Although information systems (IS) scholars have been applying neurophysiological tools for decades, a renewed call for drawing on the brain sciences as a reference discipline for the IS field took place in December 2007, at the International Conference on Information Systems (ICIS) and at two pre-ICIS meetings (see Riedl and Léger 2016, p. 73, for details on the genesis of NeuroIS). Angelika Dimoka, Paul A. Pavlou, and Fred D. Davis coined the term *NeuroIS* (Dimoka et al. 2007).[1] This concept endorses the application of knowledge and tools from neuroscience to the field of IS research. Although NeuroIS has existed for less than a decade, a number of conceptual articles (e.g., Dimoka et al. 2011, 2012; Loos et al. 2010; Riedl 2009; Riedl et al. 2010a) and empirical articles have already been published (a concise review of the empirical NeuroIS literature is provided in Appendix A).[2] Moreover, two leading IS journals, *Journal of Management Information Systems* (JMIS, Vol. 30, No. 4, 2014) and *Journal of the Association for Information Systems* (JAIS, Vol. 15, No. 10, 2014), recently published special issues focused on NeuroIS.

The establishment of NeuroIS as a subfield of the IS discipline has its basis in two main factors (Riedl et al. 2010a). The first factor is that *neuroscience tools* (e.g., functional magnetic resonance imaging—fMRI), if compared to traditional instruments (e.g., surveys), are often expected to offer more objective and unbiased measurement of IS constructs (Dimoka et al. 2011, p. 687). While the nature and degree of the objectivity of neurophysiological data is a matter of ongoing discussion in the IS literature [see, for example, a paper on NeuroIS research methodology by Riedl et al. (2014)], there is agreement among IS scholars that data

---

[1] For example, Galletta and colleagues used an "earlobe pulse meter sensor and cheek skin temperature probe" to measure the stress level of computer users (Huston et al. 1993).
[2] In addition to the existing NeuroIS literature and corresponding panel discussions (e.g., Dimoka et al. 2009, 2010; Riedl et al. 2010b), the "Gmunden Retreat on NeuroIS" conference was established in 2009 (see www.NeuroIS.org). This event taking place annually in Gmunden, Austria, is exclusively focused on the advancement of the NeuroIS field.

© Springer International Publishing AG 2017
R. Riedl et al., *Neuroscience in Information Systems Research*,
Lecture Notes in Information Systems and Organisation 21,
DOI 10.1007/978-3-319-48755-7_1

captured through use of neuroscience tools help to "triangulate different measurement methods and data sources, and thereby strengthen the robustness of empirical IS studies" (Dimoka et al. 2011, p. 688). From a knowledge perspective, therefore, the use of multiple data sources (e.g., physiological, perceptual, and behavioral) to study a specific phenomenon is the most desirable approach (for an example, see Venkatraman et al. 2015).[3]

The establishment of NeuroIS is also associated with a second factor. As indicated by Dimoka et al. (2007), and later stressed in other articles (Dimoka et al. 2011, 2012; Loos et al. 2010; Riedl et al. 2010a; vom Brocke et al. 2013), IS scholars need not necessarily use neuroscience tools in order to pursue NeuroIS research. Rather, the consideration of existing *neuroscience knowledge* in IS studies is also an important part of NeuroIS research. In this context, Riedl et al. (2010a, p. 249) write: "While the usage of neuroscience tools requires access to research facilities (e.g., fMRI scanners), the application of neuroscience theories and findings is more straightforward. Therefore, benefiting from NeuroIS research does not necessarily imply conducting empirical neuroscience studies … Rather, it is equally important to apply the knowledge that has already accumulated in the neuroscience literature to inform IS research questions."

Because the focus of NeuroIS literature has been on neuroscience tools rather than on neuroscience knowledge (Dimoka 2012; Dimoka et al. 2012; Gefen et al. 2014; Hubert et al. 2017; Müller-Putz et al. 2015; Riedl et al. 2010a, 2014), and noting that not a single IS paper advocating the use of neuroscience knowledge focuses on describing the exact nature of this knowledge application process (see Appendix B, which summarizes major statements from NeuroIS papers), the present book seeks to close this gap. Using concrete research examples from the domain of online trust, we show how, based on neuroscience knowledge, IS scholars can (i) reinterpret existing behavioral findings, (ii) develop new hypotheses and, at least in some cases, (iii) test the hypotheses with non-neuroscience tools (e.g., rating scales). Thus, this book's objective is to show how IS scholars can effectively apply neuroscience knowledge in ways that do not require neuroscience tools.

Importantly, the approach described in this book must be considered as a complement to the application of neuroscience tools (as discussed in papers such as Dimoka et al. 2011, 2012; or Riedl et al. 2010a). The book, accordingly, must not be interpreted as opposing the application of neuroscience tools. However, not every IS researcher has access to neuroscience tools—particularly devices that are used for measurement (e.g., fMRI) or manipulation (e.g., Transcranial Magnetic Stimulation, TMS) of physiological activity in the brain. Accordingly, the intent of this book is to raise awareness among IS scholars for the various possibilities that neuroscience knowledge itself offers, both for IS theorizing and for testing hypotheses based only on non-neuroscience tools. Overall, a primary conclusion of

---

[3]However, multiple reasons exist that may prevent access to all data sources (e.g., costs, or specific phenomena that are not accessible through introspection).

this study is that neuroscience knowledge makes possible a deeper understanding of IS phenomena by connecting the behavioral and neural levels of analysis.

The remainder of this book is organized as follows: In Chapter "Knowledge Production in Cognitive Neuroscience: Tests of Association, Necessity, and Sufficiency" we briefly outline major knowledge production processes in cognitive neuroscience, thus developing a conceptual foundation for subsequent sections. In Chapter "Applying Knowledge of Brain Functionality without Neuroscience Tools: Three Example Studies and Abstraction of the Underlying Logic", we show IS scholars that they can better understand IS phenomena by applying knowledge of the brain, and that they can do this without the use of neuroscience tools. To illustrate how IS can effectively use neuroscience, we detail three concrete IS research examples from the domain of online trust. Drawing on these three examples and on the logic of abstraction, we formalize our approach to show that future IS research can smoothly and effectively integrate neuroscience knowledge, independent of the specific area of IS research. In Chapter "Notes on the Application of the Approach" we describe important notes on the application of the approach presented in this paper, and Chapter "Conclusion" provides concluding comments.

References

Dimoka, A. (2012). How to conduct a functional magnetic resonance (fMRI) study in social science research. *MIS Quarterly, 36*(3), 811–840.

Dimoka, A., Bagozzi, R., Banker, R., Brynjolfsson, E., Davis, F., Gupta, A., et al. (2009). NeuroIS: Hype or Hope? *ICIS 2009.*

Dimoka, A., Banker, R. D., Benbasat, I., Davis, F. D., Dennis, A. R., Gefen, D., et al. (2012). On the use of neurophysiological tools in IS research: Developing a research agenda for NeuroIS. *MIS Quarterly, 36*(3), 679–702.

Dimoka, A., Benbasat, I., Lim, K., Straub, D., & Walden, E. (2010). NeuroIS: Challenges and solutions. *ICIS 2010.*

Dimoka, A., Pavlou, P. A., & Davis, F. D. (2007). NEURO-IS: The potential of cognitive neuroscience for information systems research. *ICIS 2007.*

Dimoka, A., Pavlou, P. A., & Davis, F. D. (2011). NeuroIS: The potential of cognitive neuroscience for information systems research. *Information Systems Research, 22*(4), 687–702.

Gefen, D., Ayaz, H., & Onaral, B. (2014). Applying functional near infrared (fNIR) spectroscopy to enhance MIS research. *AIS Transactions on Human-Computer Interaction, 6*(3), 55–73.

Hubert, M., Linzmajer, M., Riedl, R., Kenning, P., Weber, B. (2017). The use of psycho-physiological interaction analysis with fMRI-data in IS research: A guideline. *Communications of the Association for Information Systems,* forthcoming.

Huston, T. L., Galletta, D. F., & Huston, J. L. (1993). The effects of computer monitoring on employee performance and stress: Results of two experimental studies. In *HICSS 1993* (pp. 568–574).

Loos, P., Riedl, R., Müller-Putz, G. R., vom Brocke, J., Davis, F. D., Banker, R., et al. (2010, December). NeuroIS: Neuroscientific approaches in the investigation and development of information systems. *Business & Information Systems Engineering, 2*(6), 395–401.

Müller-Putz, G. R., Riedl, R., & Wriessnegger, S. C. (2015). Electroencephalography (EEG) as a research tool in the information systems discipline: Foundations, measurement, and applications. *Communications of the Association for Information Systems, 37,* 911–948.

Riedl, R. (2009). Zum Erkenntnispotenzial der kognitiven Neurowissenschaften für die Wirtschaftsinformatik: Überlegungen anhand exemplarischer Anwendungen. *NeuroPsycho Economics*, *4*, 32–44.

Riedl, R., Banker, R. D., Benbasat, I., Davis, F. D., Dennis, A. R., Dimoka, A., et al. (2010a). On the foundations of NeuroIS: Reflections on the Gmunden Retreat 2009. *Communications of the Association for Information Systems, 27*(15), 243–264.

Riedl, R., Davis, F. D., & Hevner, A. R. (2014). Towards a NeuroIS research methodology: Intensifying the discussion on methods, tools, and measurement. *Journal of the Association for Information Systems, 15*(10), Article 4.

Riedl, R., & Léger, P.-M. (2016). Fundamentals of neuroIS: Information systems and the brain. Springer, Berlin.

Riedl, R., Randolph, A. B., vom Brocke, J., Léger, P.-M., & Dimoka, A. (2010b, December 12). The potential of neuroscience for human-computer interaction research. In *Proceedings of the 9th Annual Workshop on HCI Research in MIS*, St. Louis, Missouri (pp. 1–5).

Venkatraman, V., Dimoka, A., Pavlou, P. A., Vo, K., Hampton, W., Bollinger, B., et al. (2015). Predicting advertising success beyond traditional measures: New insights from neurophysiological methods and market response modeling. *Journal of Marketing Research, LII*, 436–452.

vom Brocke, J., Riedl, R., Léger, & P.-M. (2013). Application strategies for neuroscience in information systems design science research. *Journal of Computer Information Systems, 53*(3), 1–13.

Chapter 2
Knowledge Production in Cognitive Neuroscience: Tests of Association, Necessity, and Sufficiency

While all domains in neuroscience might be relevant for NeuroIS research to some degree, the field of cognitive neuroscience has been identified as the major reference discipline (e.g., Dimoka et al. 2011). Cognitive neuroscience seeks to understand "how the brain works, how its structure and function affect behavior, and ultimately how the brain enables the mind" (Gazzaniga et al. 2009, p. 2).[1] In order to develop a conceptual basis for the sections to follow, we briefly discuss how cognitive neuroscience knowledge is typically produced. Appendix C provides additional information on brain functioning from a cognitive neuroscience perspective.

A central objective of cognitive neuroscience studies is to determine how a particular mental process is implemented neurologically, and to do so by identifying the regions of the brain that are involved in a specific task. Such research often relies on fMRI, which is the most significant tool in cognitive neuroscience (Gonsalves and Cohen 2010; Poldrack 2006). Indeed, Logothetis (2008) states that fMRI "is the most important imaging advance since the introduction of X-rays by Conrad Röntgen in 1895" (p. 869), and its prominence in cognitive neuroscience research bears out that view. Given the role of fMRI in the field, and supported by fMRI guidelines that Dimoka (2012) presents in *MIS Quarterly*, along with related methodological contributions (Hubert et al. 2012, 2017; Riedl et al. 2014), we will herein briefly illustrate the logic of fMRI research.[2]

---

[1]Neuroscience is the scientific examination of the nervous system. While this discipline was a sub-discipline of biology in former times, it is an independent discipline today. As described in detail by leading academic societies such as the *Society for Neuroscience* or the *International Brain Research Organization*, neuroscience research refers to different levels of analysis (e.g., molecular, cellular, structural, or functional) and domains (e.g., evolutionary, medical, developmental, computational, or cognitive). As a function of both level of analysis and of domain, neuroscientists apply different research tools.

[2]Dimoka et al. (2012), Riedl et al. (2010), and Riedl and Léger (2016) discuss additional cognitive neuroscience and neurophysiological tools.

© Springer International Publishing AG 2017
R. Riedl et al., *Neuroscience in Information Systems Research*,
Lecture Notes in Information Systems and Organisation 21,
DOI 10.1007/978-3-319-48755-7_2

A useful example is a situation in which a cognitive neuroscientist wants to identify the brain region(s) underlying the mental process of *disgust*. After forming a clear definition of disgust (for example, "profound dislike or annoyance caused by something sickening or offensive", American Heritage Dictionary), a researcher needs to transform the concept of disgust into an experimental paradigm. Stimuli and tasks are major components of such a paradigm.[3] Disgust is one of the six basic human facial expressions representing emotional states; the others are anger, fear, happiness, sadness, and surprise (e.g., Gazzaniga et al. 2009). Thus, in order to identify the brain regions underlying the perception and processing of disgust, a researcher could present a range of photos of human faces to participants whose brains are being scanned, including in the photo range both images of faces expressing a disgust response, and control images of faces that convey other emotional states or no emotional state at all (i.e., neutral faces). Brain research of this design (e.g., Phillips et al. 1997) has, in fact, found that faces expressing disgust activate the insular cortex more strongly than do faces expressing other emotions (e.g., fear) or no emotion.

In contrast to classic neuroscience research (which is usually interested in brain data alone), cognitive neuroscience research is focused on both behavior and in underlying brain mechanisms. The design of fMRI experiments reflects this difference. For fMRI experiments in classic neuroscience, subjects are typically presented with visual stimuli (though there may be other stimuli such as auditory cues), and the scanner measures brain activity. For fMRI experiments in cognitive neuroscience, however, another element is added—subjects are usually presented with stimuli while the scanner measures brain activity, but participants are also required to state a behavioral response after the presentation of stimuli (e.g., a disgust evaluation using a Likert-type scale after presentation of faces showing varying levels of disgust). Based on the collection of both brain and behavioral data, the relationship between both data sets can be assessed, in this way using brain activity as a mediator between the stimulus perception and task execution, and a behavioral response.

Figure 2.1 (panel A) conceptually illustrates the transformation of a mental process (e.g., disgust) into stimuli (e.g., faces expressing disgust) and control stimuli (e.g., faces expressing other emotional states or no emotion). By contrasting the brain images acquired during the various experimental conditions (e.g., disgust vs. other emotional states, disgust vs. no emotion), it is possible to determine the brain regions associated with the investigated mental process. Brain research has found that the insular cortex plays a crucial role in the implementation of disgust (Gazzaniga et al. 2009, p. 383). The inference from a mental process to brain activity (i.e., from the corresponding stimulus and task to brain activity) is referred to as *forward inference* (Henson 2006).[4]

---

[3]The website http://www.cognitiveatlas.org/ describes several hundred tasks that are used in cognitive science, many of which are also used in cognitive neuroscience research.

[4]Reverse inference (i.e., an inference from neuroimaging data reported in the literature to a mental process, Poldrack 2006) is detailed in Appendix C.

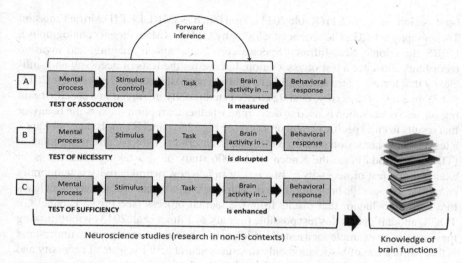

Fig. 2.1 Process of cognitive neuroscience knowledge production

Based on the logic described in Fig. 2.1 (panel A), neuroscience research has developed an impressive body of knowledge. In particular, the research has produced extensive insight into the relationship between mental processes and brain regions. As a result, an extensive knowledge base on brain functions exists today (Fig. 2.1, right), and this literature can be used by IS scholars without necessarily using neuroscience tools. In part, this knowledge is documented in online databases such as www.cognitiveatlas.org *or* www.neurosynth.org.

In our example, we indicate that activity in the insular cortex is associated with the specific mental process of disgust. However, other studies unrelated to disgust have also revealed activity in this particular region of the brain. For example, research has found that in an economic game (task), an unfair offer (stimulus) from one player (in contrast to a fair offer), elicits stronger activity in the insular cortex of the opponent's brain (Sanfey et al. 2003). Consequently, insular cortex activity is associated with unfairness. In addition to this result, a stream of research on the brain mechanisms underlying economic decision-making has found that insular cortex activity is correlated with risk (e.g., Clark et al. 2008; Mohr et al. 2010; Preuschoff et al. 2008). The cognitive neuroscience literature, therefore, provides evidence that insular cortex activity is associated with a number of mental processes such as disgust, unfairness, and risk. More generally, evidence shows that a brain region is typically activated by several mental processes (i.e., corresponding stimuli and tasks), and a single mental process (i.e., corresponding stimuli and tasks) often activates more than one brain region (e.g., Price and Friston 2005).

From an IS perspective, it is critical to understand that most knowledge that is currently documented in the cognitive neuroscience literature has been developed through *tests of association* (i.e., a mental process *correlates* with activity in a specific brain region, see Fig. 2.1) (Kable 2011). Tools such as fMRI, PET (Positron Emission Tomography), EEG (Electroencephalography), MEG (Magnetoencephalography), fNIRS (functional Near-Infrared Spectroscopy), anatomical imaging, and invasive recordings allow for a test of association, but not for the tests of necessity and sufficiency that are used less frequently in cognitive neuroscience research (Kable 2011, p. 67). In a *test of necessity*, neural activity is temporally *disrupted* in a specific brain region, and observation is used to determine whether disruption impairs the behavior that results from a specific mental process (see Fig. 2.1, panel B); tools that make such a test possible are lesion studies, TMS, and Transcranial Direct-Current Stimulation (TDCS) (cathodal) (see the Knoch et al. 2006 study on risk-taking behavior as an example of a test of necessity).[5] In a *test of sufficiency*, neural activity is temporally *enhanced* in a specific brain region in order to observe whether the enhancement leads to a specific behavior that results from the mental process (see Fig. 2.1, panel C); TDCS (anodal) makes this test possible (a study by Filmer et al. 2013 on multitasking may serve as an example for a test of sufficiency). Thus, the central point to understand is that while a test of association only measures neural activity, tests of necessity and sufficiency manipulate the neural activity. Ideally, the functional role of a brain region is established on the basis of a multi-method approach.

After a brief outline of the major processes that cognitive neuroscientists use to generate knowledge about brain functions, we demonstrate how that neuroscience knowledge can be applied in IS research without requiring neurobiological measurement.

References

Clark, L., Bechara, A., Damasio, H., Aitken, M. R. F., Sahakian, B. J., & Robbins, T. W. (2008, May). Differential effects of insular and ventromedial prefrontal cortex lesions on risky decision-making. *Brain, 131*(5), 1311–1322.

Dimoka, A. (2012). How to conduct a functional magnetic resonance (fMRI) study in social science research. *MIS Quarterly, 36*(3), 811–840.

Dimoka, A., Pavlou, P. A., & Davis, F. D. (2011, December). NeuroIS: The potential of cognitive neuroscience for information systems research. *Information Systems Research, 22*(4), 687–702.

Filmer, H. L., Mattingley, J. B., & Dux, P. E. (2013). Improved multitasking following prefrontal tDCS. *Cortex, 49*, 2845–2852.

Gazzaniga, M. S., Ivry, R., & Mangun, G. R. (2009). *Cognitive neuroscience: The biology of the mind* (3rd ed.). New York: W.W. Norton.

Gonsalves, B. D., & Cohen, N. J. (2010, November). Brain imaging, cognitive processes, and brain networks. *Perspectives on Psychological Science, 5*(6), 744–752.

Henson, R. (2006, February). Forward inference using functional neuroimaging: Dissociations versus associations. *Trends in Cognitive* Sciences, *10*(2), 64–69.

Hubert, M., Linzmajer, M., Riedl, R., Kenning, P., & Hubert, M. (2012). Introducing connectivity analysis to NeuroIS research. *ICIS 2012*.

---

[5]A lesion is a localized area of damage in the brain. Thus, a lesion can be considered a "permanent disruption" of the brain region of concern.

Hubert, M., Linzmajer, M., Riedl, R., Kenning, P., Weber, B. (2017). The use of psycho-physiological interaction analysis with fMRI-data in IS research: A guideline. *Communications of the Association for Information Systems*, forthcoming.

Kable, J. W. (2011). The cognitive neuroscience toolkit for the neuroeconomist: A functional overview. *Journal of Neuroscience, Psychology, and Economics, 4*(2), 63–84.

Knoch, D., Gianotti, L. R. R., Pascual-Leone, A., Treyer, V., Regard, M., Hohmann, M., et al. (2006). Disruption of right prefrontal cortex by low-frequency repetitive transcranial magnetic stimulation induces risk-taking behavior. *Journal of Neuroscience, 26*(24), 6469–6472.

Logothetis, N. K. (2008, June). What we can do and what we cannot do with fMRI. *Nature, 453* (7197), 869–878.

Mohr, P. N. C., Biele, G., & Heekeren, H. R. (2010, May). Neural processing of risk. *Journal of Neuroscience, 30*(19), 6613–6619.

Phillips, M. L., Young, A. W., Senior, C., Brammer, M., Andrew, C., Calder, A. J., et al. (1997, October). A specific neural substrate for perceiving facial expressions of disgust. *Nature, 389* (6650), 495–498.

Poldrack, R. A. (2006, February). Can cognitive processes be inferred from neuroimaging data? *Trends in Cognitive Sciences, 10*(2), 59–63.

Preuschoff, K., Quartz, S. R., & Bossaerts, P. (2008, March). Human insula activation reflects risk prediction errors as well as risk. *Journal of* Neuroscience, 28(11), 2745–2752.

Price, C. J., & Friston, K. J. (2005). Functional ontologies for cognition: The systematic definition of structure and function. *Cognitive Psychology, 22*(3–4), 262–275.

Riedl, R., Banker, R. D., Benbasat, I., Davis, F. D., Dennis, A. R., Dimoka, A., et al. (2010). On the foundations of NeuroIS: Reflections on the Gmunden Retreat 2009. *Communications of the Association for Information Systems, 27*(15), 243–264.

Riedl, R., Davis, F. D., & Hevner, A. R. (2014). Towards a NeuroIS research methodology: Intensifying the discussion on methods, tools, and measurement. *Journal of the Association for Information Systems, 15*(10), Article 4.

Riedl, R., & Léger, P.-M. (2016). Fundamentals of neuroIS: Information systems and the brain. Springer, Berlin.

Sanfey, A. G., Rilling, J. K., Aronson, J. A., Nystrom, L. E., & Cohen, J. D. (2003, June). The neural basis of economic decision-making in the ultimatum game. *Science, 300*(5626), 1755–1758.

Chapter 3
Applying Knowledge of Brain Functionality Without Neuroscience Tools: Three Example Studies and Abstraction of the Underlying Logic

The main question addressed in this book is how IS scholars can apply neuroscience knowledge to advance IS research without necessarily using neuroscience tools. As illustrated in Fig. 3.1, the process begins with a literature review. IS scholars must acquire knowledge of brain functions within their working area, and such knowledge is typically developed in non-IS contexts and hence is mainly documented in neuroscience journal publications, major textbooks (see, for example, Gazzaniga et al. 2013), and cognitive neuroscience databases (e.g., *Cognitive Atlas* or *Neurosynth*). IS scholars should not overlook evidence of a brain area's functional role having been established through tests of association alone, or whether the evidence shows that tests of necessity and/or tests of sufficiency are available as well (see Fig. 2.1); the more evidence from all three test domains that is available, the more the knowledge base used in the IS research is substantiated.

An example of the process is useful. IS trust researchers, for instance, should study the literature on neurobiological foundations of trust and related sub-processes such as reward, uncertainty, or mentalizing, as well as literature on abstract theories of brain functioning, such as predictive coding (see Appendix D and Sect. 3.3 for details). IS scholars could then reinterpret existing behavioral findings in the IS literature using the perspectives gained from the new neuroscience knowledge. Also, it is possible that, based on the neuroscience knowledge, IS scholars will develop new hypotheses—hypotheses that, at least in some research domains, could be tested using non-neuroscience tools (e.g., rating scales). These possibilities for IS research are summarized in Fig. 3.1.

Though not the specific focus of this paper, Fig. 3.1 illustrates that the hypotheses *can* also be tested using neuroscience tools. This is an important point because it emphasizes that the approach to gaining neuroscience information without the tools, as we explore in this book, is a complement to the application of neuroscience tools, and not a substitute. Figure 3.1 also shows that both reinterpretation of existing behavioral findings and development of new hypotheses constitute conceptual/theoretical work (the former with a backward focus on past findings, and the latter looking forward to new discoveries), while testing hypotheses through both

© Springer International Publishing AG 2017
R. Riedl et al., *Neuroscience in Information Systems Research*,
Lecture Notes in Information Systems and Organisation 21,
DOI 10.1007/978-3-319-48755-7_3

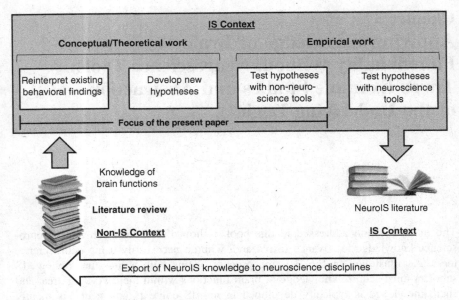

Fig. 3.1 Model of use of neuroscience knowledge in information systems research

non-neuroscience and neuroscience tools constitutes empirical work. Finally, Fig. 3.1 shows that, ideally, knowledge documented in the NeuroIS literature is also used in neuroscience disciplines.[1]

In the sections to follow we elaborate on the model in Fig. 3.1 that is based on three concrete IS research examples from the *online trust* domain. Our discussion draws from research published in three well-known academic journals:

- "Exploring Human Images in Website Design: A Multi-Method Approach." D. Cyr, M. Head, H. Larios, and B. Pan, *MIS Quarterly* (2009).
- "The Nature and Role of Feedback Text Comments in Online Marketplaces: Implications for Trust Building, Price Premiums, and Seller Differentiation." P. A. Pavlou and A. Dimoka, *Information Systems Research* (2006).
- "Online Consumer Trust and Live Help Interfaces: The Effects of Text-to-Speech Voice and Three-Dimensional Avatars," L.Y. Qiu and Izak Benbasat, *International Journal of Human-Computer Interaction* (2005).

---

[1]Whether it will be the case to a notable extent in the future cannot be predicted. If it does, however, the IS discipline can be expected to not only import knowledge from neuroscience, but also to export knowledge, and thereby to serve as a reference discipline. Consistent with Fig. 3.1's connecting arrow between the IS context and the non-IS context, Dimoka et al. (2012) wrote in their research agenda paper that NeuroIS researchers are potential "diligent contributors to the rapidly expanding neuroscience literature" (p. 696).

Appendix D describes why we have chosen the topic of online trust. We also discuss the structure of an online trust situation and a conceptual framework for trust in online environments. Importantly, we show that trust can be broken down into four major constructs (sub-processes), namely reward, uncertainty, mentalizing, and learning, and we summarize selected cognitive neuroscience knowledge of the neural foundations of these four constructs.[2] Altogether, Appendix D serves as a conceptual basis for our discussion of the following three example studies.

3.1 The Cyr et al. (2009) Study

Pictures (human images, particularly) might have influence on trust and its sub-processes in online environments. More than fifty years ago Secord (1958) proposed that one of the mechanisms for forming impressions about an external stimulus, in particular about other humans, is attributing a momentary state to an enduring attribute. Thus, if a person who is being observed, or who is interacting with someone, expresses a particular emotion such as happiness, the emotion is considered to be an enduring attribute (e.g., being a friendly person). Applying this logic to website design, Internet firms have been using human images that express positive emotions (e.g., smiling faces) on their websites to give users a positive feeling about the company, and thereby enhance trust perceptions and/or reduce distrust perceptions.

A study by Cyr et al. (2009) investigates the ways in which Internet users perceive human images to be an important element of website design. Three experimental conditions for human images were developed (see Fig. 3.2), including human images with facial features (high-human), human images without facial features (medium-human), and a control condition with no human image (no-human). They found that human images with facial features induce participants to perceive a website as more appealing and as having warmth. Moreover, higher levels of image appeal and perceived social presence resulted in higher levels of trust.

A major strength of the Cyr et al. (2009) study is that it uses three different data collection techniques—questionnaire, interview, and eye-tracking. Using the eye-tracker, the study found that the amount of time users spent viewing the images, as well as the number of fixations on the images, were significantly higher in the medium-human condition than in the other two conditions. Cyr et al. interpreted

---

[2]The review is selective because it is beyond of the scope of the present book to provide a comprehensive discussion of the neural correlates of these four constructs. Rather, against the background of Chapter "Knowledge Production in Cognitive Neuroscience: Tests of Association, Necessity, and Sufficiency" (see Fig. 2.1), it is our goal to show the diversity of stimuli and tasks used to identify the major neural correlates of the constructs. In Appendix D, however, we also provide seminal references that give comprehensive and detailed insight into the neural correlates of each construct.

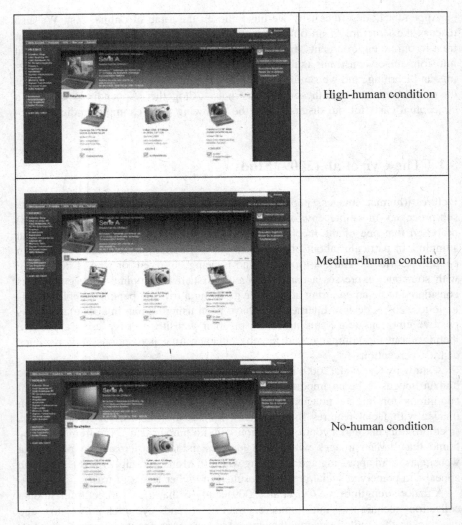

High-human condition

Medium-human condition

No-human condition

Fig. 3.2 Examples of user interfaces (*Source* Cyr et al. 2009, A2)

this result as an "unexpected finding" (p. 554). The cognitive neuroscience literature provides an explanation for this result.

When exposed to a known environment, humans typically retrieve their previous experience with that environment. If the perceived environment meets expectations, trust is likely to emerge, and if the expectations are not met, distrust is likely to emerge (Lewicki et al. 1998; Schul et al. 2004, 2008). In this context, the IS literature discusses situational normality as an important antecedent of trust (Gefen et al. 2003), which signifies a condition in which a person believes that the situation of a risky venture is in proper order and is favorable, and as a result that person expects success to be likely and failure unlikely (Baier 1986; Lewis and Weigert 1985). An Internet shopper who perceives a high degree of situational normality would believe that the online environment is appropriate, fitting, and favorable for engaging in a transaction. Thus, situational normality assures people that the Internet environment is as it ought to be, and that a shared understanding of what is happening exists (McKnight and Chervany 2001; McKnight et al. 1998; Zucker 1986).

In the Cyr et al. (2009, p. 546) study, participants reported, on average, two years online shopping experience. Moreover, they reported spending an average of fifteen hours per week online. A major characteristic of Cyr et al.'s sample, then, is that participants have considerable Internet and online shopping experience. Thus, it is likely that the participants retrieved these experiences during the experiment (see *Learning* in our conceptual framework in Appendix D).

Often, e-commerce websites either use human images with facial features or they do not present human images at all (Riegelsberger et al. 2003). The use of human images without facial features, therefore, is comparatively uncommon (see the middle column in Fig. 3.2 as an example). Considering this, we hypothesize that participants in the Cyr et al. (2009) study could have perceived the website in the medium-human condition as abnormal. Hence, the degree of perceived uncertainty was comparatively high in the medium-human condition.

This high degree of perceived abnormality and uncertainty is typically associated with activity in limbic structures, especially in the amygdala. In particular, this fact holds true for situations that involve human face perception. Brain imaging evidence indicates that the amygdala is significantly activated when people view untrustworthy faces (Winston et al. 2002) and faces that show fear (Morris et al. 1996). Moreover, medical case reports have shown that patients with complete bilateral amygdala damage judged other people to appear more trustworthy and more approachable than did people with no brain damage or patients with brain damage in other areas, providing additional evidence that the amygdala is associated with uncertainty perception (Adolphs et al. 1998, 2005).[3]

It has been shown that uncertainty, in contrast with reward, more strongly affects attention (Carretié et al. 2004; Fiorillo et al. 2003). One explanation for this phenomenon is that such an attention-enhancing mechanism serves the function of securing survival (Phelps and LaBar 2006). With respect to the findings reported by Cyr et al. (2009), brain research findings suggest that the amygdala "detects" situational abnormality, likely outside the awareness system (Davis and Whalen 2001), and through feedback to the visual cortex-enhanced attentional processes, because the amygdala has reciprocal connections with the visual cortex (Amaral et al. 1992, 2003).[4] Hence, neuroscience research has established the amygdala's function as an "early warning system" that, according to Phelps and LaBar (2006), "enables the

[3]This is an example for a test of necessity (see Fig. 2.1, panel B in "Knowledge Production in Cognitive Neuroscience: Tests of Association, Necessity, and Sufficiency").

[4]This example shows that a relationship among constructs (threat → attention) can be based on an anatomical (reciprocal) relationship between two brain regions (visual cortex ↔ amygdala).

quick detection of emotional, and potentially threatening, stimuli" (p. 433), and then results in *increased attention*, particularly because a state of increased attention facilitates information acquisition and processing (Fiorillo et al. 2003).

This theoretical mechanism is· in line with the data reported in the Cyr et al. (2009) study. Their research found that the amount of time participants spent viewing the images, as well as the number of fixations on the images, were significantly higher in the medium-human condition compared to the other two conditions, and that there was no significant difference between the high-human and no-human conditions regarding both viewing time and fixations (see in the Cyr et al. study, p. 553, Figs. 3 and 4, as well as Table 9). Based on the findings from brain research presented, we *hypothesize* that the perception of situational abnormality in the medium-human condition caused the participants to increase their attention (longer viewing time) and to search for more information (more fixations) in order to reduce their uncertainty perceptions. Figure 3.3 summarizes how neuroscience discoveries are used to explain the original finding of the Cyr et al. (2009) study. The original Cyr et al. finding is illustrated in panel A, and the theorizing based on neuroscience knowledge is illustrated in panel B.

Interestingly, the interview data from the Cyr et al. study reveals that participants perceived the medium-human condition as "odd" and "distracting." It follows that non-neuroscience tools (here, the interview) enables a test of the hypothesis that participant attention increases (a longer viewing time) in the medium-human condition in order to reduce uncertainty perceptions, which were, in turn, caused by perceptions of situational abnormality. The word "odd," in particular, may be interpreted as a first indication that our situational abnormality and uncertainty explanations have merit. Future research could formally test this hypothesis by using

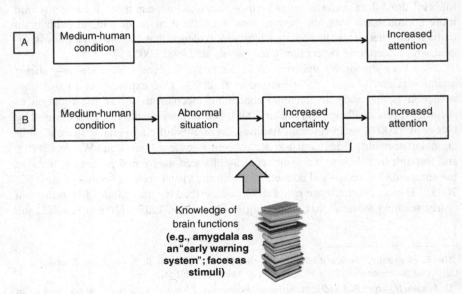

Fig. 3.3 Possible explanation for a finding reported in Cyr et al. (2009) based on neuroscience knowledge

multi-item survey instruments that are designed to measure perceptions of situational abnormality and uncertainty.[5]

Altogether, the example of the Cyr et al. (2009) study shows that neuroscience knowledge can be used to reinterpret behavioral findings—here, findings that were characterized as "unexpected" (p. 554) in the original study. Moreover, on the basis of that neuroscience knowledge, a new hypothesis was formulated (see the mediation process in Fig. 3.3, panel B), which can be tested in future research that is based on non-neuroscience tools (here, the survey).[6]

3.2 The Pavlou and Dimoka (2006) Study

A research agenda by Gefen et al. (2008, p. 282) identifies "trust building with text" as an important IS research topic. We share this view, because online reputation mechanisms—which may considerably influence the perceived trustworthiness of an online seller (Zacharia and Maes 2000)—are typically based on textual feedback posted by buyers who evaluate the quality of the transaction they have conducted with a seller.[7]

Pavlou and Dimoka's (2006) study investigated over 10,000 feedback text comments in eBay's online auction marketplace, and found that the addition of text comments to numerical ratings helps to explain a greater variance in price premiums. Hence, the study allows for a better understanding of the success of online marketplaces that rely primarily on text comments (versus crude numerical ratings) to differentiate among sellers and to prevent a market of lemon sellers (Akerlof 1970). Specifically, the study used both outstanding and abysmal text comments from eBay sellers' feedback profiles to investigate the effects on benevolence and credibility perceptions of buyers (see Table 3.1). Results indicated that outstanding benevolence and credibility comments positively affect benevolence and credibility perceptions, whereas abysmal benevolence and credibility comments negatively affect benevolence and credibility perceptions. Finally, both benevolence and credibility perceptions were shown to positively affect sellers' pricing power.

Although it was not the objective of this study to explain the results based on reward and uncertainty perceptions (i.e., positive and negative emotions), we

---

[5]For example, Pavlou et al. (2007) used a 4-item instrument to measure perceived uncertainty (see p. 135), and Moody et al. (2014) developed an 8-item instrument to measure situational abnormality (see p. 280). IS researchers could use (modified versions of) these instruments.

[6]As indicated in Fig. 2.1 (right), scholars who have access to functional brain imaging could also test the hypothesis on a brain level, with the visual cortex and the amygdala as the main regions of interest.

[7]For example, one study (Ghose et al. 2005) investigated data from Amazon's secondary online marketplace and found that having adjective-noun pairs such as "wonderful/product," "perfect/transaction," or "fast/shipping" in the customer feedback profile (a specific form of online reputation mechanism) imposes an economically significant impact on a seller's pricing power.

Table 3.1 Examples of outstanding and abysmal text comments from eBay feedback profiles

Benevolence		Credibility	
Outstanding	Abysmal	Outstanding	Abysmal
Seller went above and beyond her duty to help me. She had a solution to every problem! I am indebted to her.	Seller collects payment and does not send expensive items. Buyer beware!	Extremely prompt seller. I was thrilled with the speed of the service I received.	Very displeased with such incompetence and negligence
Seller was really tolerant and did not take advantage of my bidding error	Product's condition profoundly misrepresented; this is a copied CD, not original; beware!!!	Super-fast transaction and delivery. Excellent seller!	Overnight shipping took two weeks! Useless seller …

Source Pavlou and Dimoka (2006, pp. 401–402)

nonetheless hypothesize, based on brain research evidence (described below), that reward and uncertainty perceptions mediated the positive and negative effects of the text comments on trust perceptions (benevolence and credibility) and price premiums. Table 3.1 indicates examples of outstanding and abysmal benevolence and credibility comments as reported in the Pavlou and Dimoka study.

Written communication, which emerged approximately 5000 years ago (Kock 2009), contributed to the successful development of the human species. One factor that favored the importance of written communication for the prosperous development of human society is the fact that humans are able to authentically communicate emotions through this modality. Hence, although body language, facial expression, and pitch of the voice make possible an authentic communication of emotions in face-to-face settings, there is no doubt that written communication, also, allows for the effective transfer of emotions (Aman and Szpakowicz 2007).

Cognitive neuroscience research has extensively studied the neural effects of reading of emotional text. Recent studies in this research domain usually draw upon the concept of "neural reuse" (Anderson 2010). According to this concept, neural circuits established for one purpose during evolution or normal development are reused (i.e., exploited, recycled, and/or redeployed) for different uses, yet reuse does not lead to loss of the original function. It follows that phylogenetically younger processes (here, reading of emotional text), rely heavily on existing emotion-processing regions (i.e., areas involved in processing affective information inherent in other stimuli such as pictures, faces, or odor), and these regions are predominantly located in the limbic system.[8] In a recent review, Briesemeister et al. (2015) summarize the state of the field as follows: "Recent studies have suggested

---

[8]According to standard psycholinguistic theory, the opposing view is that "emotional responses to words are generated within the reading network itself subsequent to semantic activation" (Ponz et al. 2014, p. 619).

that the processing of single affective words relies on emotion networks in the brain, such as the anterior and posterior cingulate cortex, the medial temporal lobe including hippocampus and parahippocampal gyrus, the amygdala, and the orbitofrontal cortex" (p. 289). Their findings confirm prior literature review results (e.g., Citron 2012). One concrete research example is a study by Ponz et al. (2014), who show that the insular cortex, a brain region related to perception of facial expressions of disgust (for details, see Chapter "Knowledge Production in Cognitive Neuroscience: Tests of Association, Necessity, and Sufficiency"), is also involved in reading disgust-related words (e.g., infection, vomit, and vermin).

In addition to describing emotions as discrete states that vary along a continuum (for example, disgust in the Ponz et al. study), another way to approach the categorization of emotions is by using a 2 × 2 matrix with the factors valence (pleasant—unpleasant, or good—bad) and arousal (the intensity of the emotional response, high—low). Moreover, it is possible to categorize different emotional reactions on the basis of the motivation to exhibit either approach or avoidance behavior. Based on the 2 × 2 emotion categorization, Lewis et al.'s (2007) seminal study investigating separate domains of positive and negative words demonstrates that increasing arousal for positive words enhances activity within the striatum, a brain area related to reward processing and positive emotion (e.g., Schultz 2006), while increasing arousal for negative words engages the brainstem, amygdala, and insula—brain areas that have been shown to correlate with negative emotions and, more generally, with states of high motivational and behavioral significance (e.g., Gazzaniga et al. 2009).

Overall, the cognitive neuroscience findings presented in this section allow for a reinterpretation of the behavioral results reported in Pavlou and Dimoka (2006), which makes possible a new hypothesis based on that neuroscience knowledge. The outstanding text comments in Table 3.1 can be expected to result in a pleasant feeling (e.g., "*I was thrilled with the speed of the service I received.*"). The abysmal text comments, in contrast, could have led to unpleasant feelings (e.g., "*Product's condition profoundly misrepresented; this is a copied CD, not original; beware!!!*"). Moreover, outstanding comments are likely to result in trust and approach behavior, whereas abysmal comments may lead to distrust and avoidance behavior. The level of response (pleasantness or unpleasantness), in the cases of outstanding and abysmal comments, typically strengthens trust and distrust behavior. These hypotheses can be tested in future IS research that uses traditional (non-neuroscience) tools such as a survey or an interview.[9,10]

---

[9]Drawing upon Russell (2003), we define pleasantness of a feeling as a "neurophysiological state that is *consciously accessible* as a simple, nonreflective feeling that is an integral blend of hedonic (pleasure–displeasure) and arousal (sleepy–activated) values" (p. 147, italics added). A semantic differential scale (1–7) with six items (e.g., displeased–pleased) was developed by De Guinea and Webster (2013) to capture user feelings (p. 1169). Moreover, IS researchers could capture arousal by adapting measurement instruments developed more than three decades ago to capture self-reported arousal (Cox and Mackay 1985; King et al. 1983; Mackay et al. 1978).

[10]In this context, it is important to refer to the right side of Fig. 3.1, because two fMRI studies in the NeuroIS field (Dimoka 2010; Riedl et al. 2010) use textual formulations with varying levels of

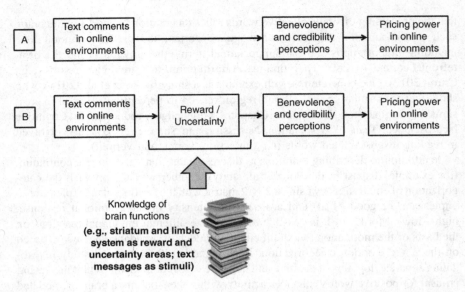

Fig. 3.4 Possible explanation for a finding reported in Pavlou and Dimoka (2006) that is based on neuroscience knowledge

Figure 3.4 summarizes how neuroscience findings can be used to explain a relationship between two constructs of the Pavlou and Dimoka (2006) study—the original finding is illustrated in panel A and the theorizing based on neuroscience knowledge is illustrated in panel B. Specifically, the example shows that neuroscience knowledge can be used to develop an ex-post explanation for the relationship between text comments in online environments (outstanding versus abysmal) and self-reported benevolence or credibility perceptions. Our hypothesis

---

(Footnote 10 continued)

trustworthiness that are embedded in websites in order to study the effects on brain activation patterns. Although both studies did not explicitly control for the emotional content inherent in the texts (with respect to both valence and arousal), the brain activation patterns clearly indicate that text in online environments activates brain circuits for both reward and uncertainty perceptions (see our conceptual framework in Appendix D). In particular, the studies found significant activation in the basal ganglia (e.g., striatum) and limbic regions (e.g., amygdala, insular cortex, orbitofrontal cortex); the basal ganglia are known to have a prominent role in reward processing, and the limbic regions are recognized for their role in the processing of negative emotions—of uncertainty, in particular (Kolb and Wishaw 2009, pp. 567–574). Thus, both the Dimoka (2010) and the Riedl et al. (2010) studies provide evidence that activation in brain areas that are related to reward and to uncertainty (i.e., positive and negative emotions) mediates the influence of textual formulations on trust and distrust perceptions, which in turn affect important outcome variables in online environments, such as a seller's pricing power. This influence of textual formulations on reward or uncertainty perceptions is usually mediated by memory, because the encoding of textual semantics is not innate, but is, rather, an ability that has to be learned (Dapretto and Bookheimer 1999).

is that reward (neurologically implemented through activation in the striatum, the consequence of perception of outstanding text comments in online feedback profiles) and uncertainty (neurologically implemented through activation in the limbic region, predominantly in the amygdala, insula, and orbitofrontal cortex, the consequence of abysmal text comments) mediate users' benevolence and credibility ratings, which in turn affect a seller's pricing power.

3.3 The Qiu and Benbasat (2005) Study

We have thus far discussed human images and textual formulations on websites in the context of online trust. In the following, we discuss research that investigates the trust-building potential of avatars, here defined as computer-generated visual representations of humans (Nowak and Rauh 2005, p. 153). Foremost, the use of avatars on websites, in contrast to text-based interfaces, may draw the attention of the user, creating a higher level of engagement (Bente et al. 2008). As well, however, this may distract the user from the task which the avatar is aiding (Yee et al. 2007). It is therefore not possible to claim that avatar-based interfaces are generally more advantageous than text-based interfaces.

Online shopping environments typically use avatars as virtual sales consultants (Holzwarth et al. 2006). In contrast to interaction with a website based on text alone, interaction with an avatar-based website has been shown to result in increased perceptions of closeness and connectedness, which is referred to as social presence (Bente et al. 2008). Moreover, a higher degree of social presence has been demonstrated to enhance users' trusting beliefs and, ultimately, their intentions to use an avatar as a decision aid (Qiu and Benbasat 2009).

Based on the observation that increasingly more companies have begun to provide live help interfaces (e.g., instant messaging or text chatting) on their websites to support interactions between online consumers and customer service representatives, Qiu and Benbasat (2005) conducted a laboratory experiment to investigate the effects that different design features of live help interfaces can have on consumer trust. The study investigated six experimental conditions (3 × 2 factorial design): text output versus voice output versus text plus voice output; no avatar versus 3D avatar (see Fig. 3.5, which shows examples of stimulus material). Voice output was implemented as a text-to-speech (TTS) voice, that is, a computer-based system was used to read aloud answers to a user's questions.

The findings of the study show that the presence of voice output significantly increased consumers' trust toward the customer service representative, if compared to the other two conditions (i.e., text output and text plus voice output).[11] However,

---

[11]The combination of text plus voice "may add extra burdens to users' cognitive efforts and decrease the perceived social cues from voice" (Qiu and Benbasat 2005, p. 89). This may explain why voice output alone results in higher trust than does text plus voice.

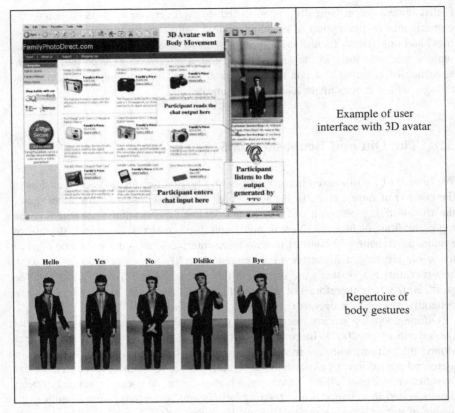

Fig. 3.5 Example of user interface with 3D avatar and repertoire of body gestures (*Source* Qiu and Benbasat 2005, pp. 84/86)

no empirical support was found for the hypothesis that 3D avatars will generate a higher level of trust in the customer service representative than will interfaces without an avatar.

This unexpected result is explained by Qiu and Benbasat (2005, p. 89): "One apparent explanation may be that people do not care what the virtual CSR (customer service representative) looks like when they are looking for help; conversely, the lack of support could be attributed to the limitations of technology. It appears that currently available Web-based animated embodiments still lack the vividness and interactivity that make a strong peripheral cue to influence users' judgments. In this study, although the 3D avatar could generate basic body language (see Fig. 3.5, bottom), *the lack of facial expressions greatly degraded the fidelity and warmness one might perceive in face-to-face communication*" (italics and reference to Fig. 3.5 in square brackets added).

Avatar-based communication can be used to increase reward perceptions and reduce uncertainty perceptions in online interaction because it more closely resembles the "natural" medium of communication, namely face-to-face. Media naturalness theory (Kock 2004, 2005, 2009) defines the naturalness of a communication medium as the degree to which the medium is similar to face-to-face communication. However, it is clear that if avatars do not possess facial expressions (or if such expressions are implemented in an unprofessional or unnatural way), they do not tap avatars' potential for enhancing reward and reducing uncertainty, which could then lead to higher levels of trust or reduced levels of distrust (see the conceptual framework in Appendix D). Rather, researchers argue that deviations from face-to-face communication, which exist in cases where facial expressions are either completely missing or are unnaturally implemented, typically result in communication ambiguity (Kock 2004, 2005, 2009).

Non-verbal communication, and facial expressions in particular, typically transfers a large amount of emotional information. To date, avatar research has investigated the effects of several avatar characteristics on emotional perception. One study (Ku et al. 2005) shows that the intensity of the happiness or anger in avatar faces produces linear changes in a subject's self-reported emotion, both with respect to valence and to arousal. Another study (Weibel et al. 2010) manipulates an avatar's pupil size and eye blink frequency to investigate whether, and how, non-verbal communication affects impression formation, which is a highly emotion-laden construct. The results indicate that avatars with large pupils and slow eye blink frequency are perceived to be more sociable and attractive, thereby resulting in positively valenced emotion. This result suggests that an avatar's facial characteristics (such as pupil size and eye blink frequency) may result in positive emotions.

Neuroimaging research provides biological evidence that makes possible a better understanding of the presented findings. The amygdala is a core structure for emotional face processing (for a brief review, see Gazzaniga et al. 2009). In one fMRI study (Moser et al. 2007), participants performed facial emotion recognition tasks that were based on human and avatar stimuli. Although the neural responses were significantly stronger for human faces in face-sensitive structures (in particular in the fusiform gyrus), robust amygdala activation was found in response to emotion in both human and avatar faces. Other studies using various techniques to capture physiological reactions to avatars and avatar-like robots (heart rate, electrodermal activity, electromyographic activity, and electroencephalography) found similar results (Dubal et al. 2011; Garau et al. 2005; Weyers et al. 2006). Considering this, research suggests that avatars have the potential to elicit strong emotional reactions in humans, but only if facial expressions are implemented in a way that makes the effective communication of emotions possible. Thus, cognitive neuroscience research offers a neurobiological explanation for Qiu and Benbasat's (2005) view that a lack of facial expressions in the avatars may have caused their non-significant results.

This explanation can be reconciled with an important neuroscience concept referred to as *predictive coding*. According to Huang and Rao (2011), predictive coding "postulates that neural networks learn the statistical regularities inherent in the natural world and reduce redundancy by removing the predictable components of the input, transmitting only what is not predictable (the residual errors in prediction)" (p. 580).[12] Thus, a reinterpretation of the Qiu and Benbasat (2005) data that is consistent with the theory of predictive coding is that when participants see the 3D avatars they *expect* human-like behavior, particularly facial expressions. However, because the 3D avatars in the Qiu and Benbasat study were not implemented in a technically optimal way (resulting in a lack of facial expressions), deviations from the expectations occurred, creating the potential for negatively affecting trust perceptions.

Most people automatically expect human-like 3D avatars to have facial expressions (see Fig. 3.5). However, people also quickly recognize a mismatch between the human-like appearance and the lack of facial expressions. This mismatch perception is based on a larger prediction error which, in turn, is typically associated with increased activity in multiple brain regions across different hierarchical levels (Friston 2010; Kitzbichler et al. 2011; Rao and Ballard 1999; Saygin et al. 2012). This fact, importantly, is substantiated by other brain function principles, particularly repetition suppression, which refers to the phenomenon of reduced neural response to a repeated stimulus (Grill-Spector et al. 2006; Henson and Rugg 2003). In essence, repeating a stimulus in the same context means that the possibility of a prediction error becomes smaller with every additional repeated stimulus.

Relating the concept of predictive coding to our framework in Fig. 3.1 (right), we stress that this explanation constitutes a new hypothesis that can be tested in future studies with the use of brain imaging tools. One example hypothesis is that a mismatch between the human-like appearance and the lack of facial expressions of avatars results in greater overall brain activity, if compared to a matching situation (i.e., human-like appearance and facial expressions). Importantly, greater physiological (brain) activity typically relates positively to factors of a self-reported mental load or fatigue (a negative perception)—a fact that holds true if a specific threshold of physiological (brain) activity is exceeded (e.g., Ishii et al. 2013; Trimmel et al. 2003; Vicente et al. 1987). Hence, the predictive coding hypothesis can even be tested with non-neuroscience tools, though indirectly (e.g., with analog rating scales).[13] The overall negative state, induced by increased mental load, might

---

[12]For detailed descriptions of predictive coding and related brain function principles, see, for example, Friston (2003, 2005) and Kilner et al. (2007).

[13]For example, in the Trimmel et al. (2003) study, after completing a human–computer interaction task based on a 163-mm analog scale, participants self-reported their mental load and it was found that users with high self-reported mental strain had significant heart rate elevations (beats per minute) during task completion. Earlier work (e.g., Vicente et al. 1987) found strong correlations between subjective ratings of mental effort and physiological measures (e.g., sinus arrhythmia). With respect to the correlation of brain activity and self-reported mental load, a recent MEG study (Ishii et al. 2013) found that Alpha band (8–13 Hz) power in task-relevant brain areas decreased after performing the mental fatigue-inducing tasks. Importantly, this decrease was positively correlated with the self-reported level of mental fatigue (based on a visual analog scale) and the level of sympathetic nerve activity.

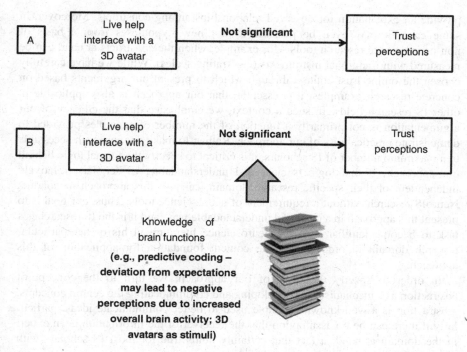

Fig. 3.6 Possible explanation for a non-significant finding reported in Qiu and Benbasat (2005) based on neuroscience knowledge

explain why Qiu and Benbasat (2005) did not find increased levels of trust in the 3D avatar condition.

Figure 3.6 summarizes how neuroscience findings can be used to explain a non-significant relationship between two constructs of the Qiu and Benbasat (2005) study. The original Qiu and Benbasat finding is illustrated in panel A, and the theorizing based on neuroscience knowledge is illustrated in panel B. Specifically, the example shows that neuroscience knowledge can be used to develop an ex-post explanation for the non-significant relationship between a 3D avatar (used as a live help interface) and trust perceptions. The hypothesis is that deviation from expectations has led to negative perceptions due to increased overall brain activity (predictive coding), and hence the live help interface with a 3D avatar did not positively affect perceived trust.

3.4 Formalizing the Logic Behind Our Argumentation

Using a base of three concrete research examples taken from the field of online trust, we have discussed that IS scholars can use neuroscience knowledge to reinterpret existing behavioral findings, and to develop new hypotheses that may provide an explanation for observed relationships among constructs. Moreover, in some cases it may even be possible to test new hypotheses from a basis of non-neuroscience research tools—for example, when the constructs at hand can be measured with traditional instruments (e.g., rating scales). While we have carefully chosen the online trust context through which to present our arguments based on concrete research examples, it is essential that our approach is also applicable in other IS research fields. In such a context, we emphasize that the efficacy of our argumentation is not primarily a function of the number of examples provided to demonstrate practicability of our approach. Rather, to make our approach accessible to a maximum number of IS scholars, it is critical to specify the formal logic behind our approach. If that logic is clear and understandable, IS scholars, relatively independent of their specific research domain, can use this approach to conduct NeuroIS research without a requirement of neuroscience tools. Thus, our goal is to present this approach in a clear and understandable way, and it is the IS researcher's task to become familiar with the neuroscience literature in his or her particular research domain in order to lay the content foundation for application of this approach.

In order to specify the logic of our approach, we turn to the concept of abstraction as a mechanism for developing and communicating the central concepts. Abstraction is a well-known principle used to better communicate ideas, particularly if there can be no assumption that the receiver of the information is an expert in the domain at hand, a fact that certainly holds true for most IS scholars with respect to their neuroscience expertise.[14]

What is abstraction? As a concept that is widely used in many scientific disciplines, we define abstraction as a process by which information about an object is filtered, thereby choosing only the aspects that are relevant for a particular purpose. Generally, abstraction is based on inductive thought processes that omit details of an object in order to derive something more general; this principle was first systematically described by Francis Bacon who wrote: "[S]low and faithful toil gathers information from things and brings it into understanding" (cited after Farrington 1964, p. 89).

Figure 3.7 summarizes the logic behind the three example papers discussed in Sects. 3.1–3.3. Panel A summarizes our discussion of the Cyr et al. (2009) paper, panels B and C focus on discussion of the Pavlou and Dimoka (2006) paper, and panel D addresses discussion of the Qiu and Benbasat (2005) paper. In Fig. 3.8, based on inductive thought processes across panels A, B, C, and D, we present our logic in a general way.

As outlined in Sect. 3.1, the Cyr et al. (2009) study focuses on faces in online environments (= IS-relevant stimulus) based on a trust evaluation task (= IS task). Also, we discuss the neuroscience research observation that human face perception

---

[14]For example, Chan et al. (1993) have shown that naïve database users' query performance (accuracy, time, confidence) was better when using higher abstraction levels of user–database interaction. This evidence suggests that abstraction is also a viable principle for effectively communicating our ideas in the present research context.

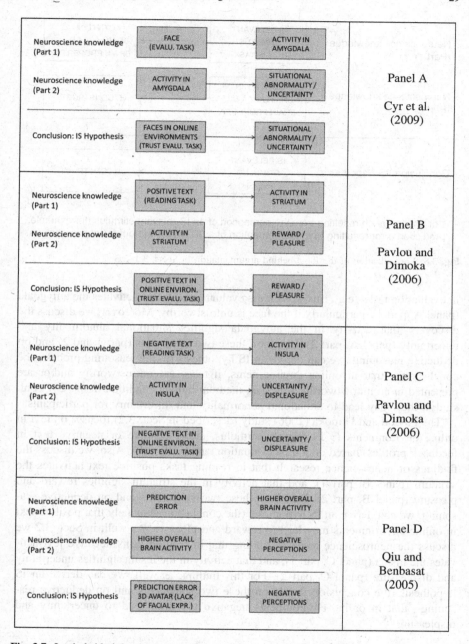

Fig. 3.7 Logic behind the three example papers discussed in Sects. 3.1–3.3

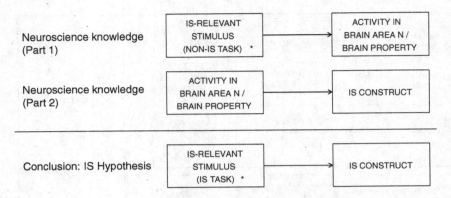

* or phenomenon resulting from the perception of an IS-relevant stimulus (for example, prediction error resulting from the perception of a 3D avatar without facial expressions)

Fig. 3.8 Formalization of the logic behind argumentation in Sects. 3.1–3.3

in evaluation tasks (e.g., trustworthiness evaluation of faces) activates the amygdala (panel A, part 1), particularly if the face is untrustworthy. Moreover, we discuss the discovery that activity in the amygdala signifies situational abnormality and uncertainty (panel A, part 2). Based on these two statements (part 1 and 2) and on deductive reasoning, we can derive an IS hypothesis (in the reasoning process, the conclusion)—that in online environments, if faces are untrustworthy and/or are presented in an untrustworthy way (i.e., medium-human condition in the Cyr et al. study), they may lead to situational abnormality and uncertainty for participants.

The Pavlou and Dimoka (2006) study (described in Sect. 3.2) focuses on text in online environments (= IS-relevant stimulus, either positive or negative text in feedback profiles) based on a trust evaluation task (= IS task). Also, we discuss the findings of neuroscience research that in reading tasks positive text activates the striatum (panel B, part 1), and that activity in the striatum signifies reward and pleasure (panel B, part 2). Based on these two statements and on deductive reasoning, we can derive an IS hypothesis (the conclusion)—namely that positive text in online environments may lead to reward and pleasure. As well, in Sect. 3.2 we discuss the neuroscience research finding that in reading tasks negative text activates the insula (panel C, part 1), and that activity in the insula signifies uncertainty and displeasure (panel C, part 2). For this finding, as well, we can derive an IS hypothesis (the conclusion) based on these two statements and on deductive reasoning—that in online environments negative text may lead to uncertainty and displeasure.[15]

---

[15]Three conclusions presented in Fig. 3.7 (panels A, B, and C) have been embedded into the causal chains described in Cyr et al. (2009) (see Fig. 3.3, panel B) and Pavlou and Dimoka (2006) (see Fig. 3.4, panel B) to provide an explanation for the observed relationships among constructs in these two example IS studies. Moreover, the conclusion presented in Fig. 3.7 (panel D) is

The Qiu and Benbasat (2005) study (described in Sect. 3.3) focuses on a live help interface with a 3D avatar (= IS-relevant stimulus) that is based on a trust evaluation task (= IS task). We discuss the neuroscience research finding that a prediction error may lead to higher overall brain activity (panel D, part 1), and that higher overall brain activity may result in negative perceptions (panel D, part 2). Based on these two statements and on deductive reasoning, we can derive an IS hypothesis (the conclusion)—that a prediction error (here, in the context of 3D avatars that lack facial expressions despite expectations of such) may lead to negative perceptions (which, in turn, may prevent trust perceptions from emerging).

Figure 3.8 presents the overall logic that we inductively generalized from our examples in panels A, B, C, and D (see Fig. 3.7). Neuroscience research that uses an IS-relevant stimulus (e.g., text), usually based on non-IS tasks (e.g., plain text reading), must be identified and read. Thereby, the IS researcher acquires knowledge (i) on the neural correlates of that stimulus and task or (ii) on more general brain properties—an example for (ii) is that a larger prediction error is associated with increased activity in multiple brain regions across different hierarchical levels (Friston 2010; Kitzbichler et al. 2011; Rao and Ballard 1999; Saygin et al. 2012) (see part 1). Next, the IS researcher has to read further neuroscience literature in order to identify (i) other functional roles of a given brain area or (ii) the effects of a given general brain property.[16]

With respect to the identification of the neural correlates of mental processes, IS researchers are advised to concentrate their research efforts on the major functions of a brain area, particularly if those functions constitute, or resemble, an IS construct.[17] Thereby, a relationship can be established between activity in a specific brain area and an IS construct (see part 2). Next, by logical connection of part 1 and part 2 (the neuroscience knowledge), an IS hypothesis can be deduced—namely, that the IS-relevant stimulus in an IS task, or a phenomenon that results from perception of that stimulus (e.g., perception of a 3D avatar without facial features may lead to a large prediction error), does affect the IS construct. This conclusion, the IS hypothesis, may help to (better) explain an observed relationship (see the examples Cyr et al. 2009 and Pavlou and Dimoka 2006), or a non-significant result (see the example Qiu and Benbasat 2005). In addition to this reinterpretation of existing behavioral findings, the IS hypothesis may also be empirically tested based on non-neuroscience tools alone (see Fig. 3.1).

(Footnote 15 continued)

illustrated as an explanation for a non-significant result in the Qiu and Benbasat (2005) study (see Fig. 3.6, panel B).

[16]As an example of an effect of a general brain property, greater overall brain activity triggered by a larger prediction error may lead to increased negative perceptions because people usually do not like levels of mental load that are too high.

[17]Here, we define an IS construct as any concept that constitutes an IS-relevant phenomenon.

References

Adolphs, R., Gosselin, F., Buchanan, T. W., Tranel, D., Schyns, P., & Damasio, A. R. (2005, January). A mechanism for impaired fear recognition after amygdala damage. *Nature, 433* (7021), 68–72.

Adolphs, R., Tranel, D., & Damasio, A. R. (1998, January). The human amygdala in social judgment. *Nature, 393*(6684), 470–474.

Akerlof, G. A. (1970, August). Market for lemons—quality uncertainty and market mechanism. *Quarterly Journal of Economics Source, 84*(3), 488–500.

Aman, S., & Szpakowicz, S. (2007). Identifying expressions of emotion in text. In V. Matousek, P. Mautner (Eds.), *Lecture Notes in Computer Science* (4629:2007). Springer, pp. 196–205.

Amaral, D. G., Behniea, H., & Kelly, J. L. (2003). Topographic organization of projections from the amygdala to the visual cortex in the macaque monkey. *Neuroscience, 118*, 1099–1120.

Amaral, D. G., Price, J. L., Pitkanen, A., & Carmichael, S. T. (1992). Anatomical organization of the primate amygdaloid complex. In J. P. Aggleton (Ed.), *The amygdala: Neurobiological aspects of emotion, memory, and mental dysfunction* (pp. 1–65). Wiley: New York.

Anderson, M. L. (2010). Neural reuse: A fundamental organizational principle of the brain. *Behavioral and Brain Sciences, 33*, 245–313.

Baier, A. (1986, January). Trust and antitrust. *Ethics, 96*(2), 231–260.

Bente, G., Rüggenberg, S., Krämer, N. C., & Eschenburg, F. (2008, April). Avatar-mediated networking: Increasing social presence and interpersonal trust in net-based collaborations. *Human Communication Research, 34*(2), 287–318.

Briesemeister, B. B., Kuchinke, L., Jacobs, A. M., & Braun, M. (2015). Emotions in reading: Dissociation of happiness and positivity. *Cognitive, Affective, & Behavioral Neuroscience, 15*(2), 287–298.

Carretié, L., Hinojosa, J. A., Martin-Loeches, M., Mercado, F., & Tapia, M. (2004, August). Automatic attention to emotional stimuli: Neural correlates. *Human Brain Mapping, 22*(4), 290–299.

Chan, H. C., Wei, K. K., & Siau, K. L. (1993). User-database interface: The effect of abstraction levels on query performance. *MIS Quarterly, 17*(4), 441–464.

Citron, F. M. M. (2012). Neural correlates of written emotion word processing: A review of recent electrophysiological and hemodynamic neuroimaging studies. *Brain and Language, 122*, 211–226.

Cox, T., & Mackay, C. (1985). The measurement of self-reported stress and arousal. *British Journal of Psychology, 76*, 183–186.

Cyr, D., Head, M., Larios, H., & Pan, B. (2009, September). Exploring human images in website design: A multi-method approach. *MIS Quarterly, 33*(3), 539–566.

Dapretto, M., & Bookheimer, S. Y. (1999, October). Form and content: Dissociating syntax and semantics in sentence comprehension. *Neuron, 24*, 427–432.

Davis, M., & Whalen, P. J. (2001). The amygdala: Vigilance and emotion. *Molecular Psychiatry, 6*, 13–34.

De Guinea, A. O., & Webster, J. (2013). An investigation of information systems use patterns: Technological events as triggers, the effect of time, and consequences for performance. *MIS Quarterly, 37*(4), 1165–1188.

Dimoka, A. (2010, June). What does the brain tell us about trust and distrust? Evidence from a functional neuroimaging study. *MIS Quarterly, 34*(2), 373–396.

Dimoka, A., Banker, R. D., Benbasat, I., Davis, F. D., Dennis, A. R., & Gefen, D., et al. (2012). On the use of neurophysiological tools in IS research: Developing a research agenda for NeuroIS. *MIS Quarterly, 36*(3), 679–702.

Dubal, S., Foucher, A., Jouvent, R., & Nadel, J. (2011, January). Human brain spots emotion in non humanoid robots. *Social Cognitive and Affective Neuroscience, 6*(1), 90–97.

Farrington, B. (1964). *The philosophy of Francis Bacon.* Liverpool: Liverpool University Press.

Fiorillo, C. D., Tobler, P. N., & Schultz W. (2003, March). Discrete coding of reward probability and uncertainty by dopamine neurons. *Science, 299*(5614), 1898–1902.

Friston, K. (2003). Learning and inference in the brain. *Neural Networks, 16*, 1325–1352.

Friston, K. (2005). A theory of cortical responses. *Philosophical Transactions of the Royal Society–B Biological Sciences, 360*, 815–836.

Friston, K. (2010). The free-energy principle: A unified brain theory? *Nature Reviews Neuroscience, 11*, 127–138.

Garau, M., Slater, M., Pertaub, D.-P., & Razzaque, S. (2005, February). The responses of people to virtual humans in an immersive virtual environment. *Presence: Teleoperators and Virtual Environments, 14*(1), 104–116.

Gazzaniga, M. S., Ivry, R., & Mangun, G. R. (2009). *Cognitive neuroscience: The biology of the mind* (3rd ed.). New York: W.W. Norton.

Gazzaniga, M. S., Ivry, R., & Mangun, G. R. (2013). *Cognitive Neuroscience: The Biology of the Mind* (4th ed.). New York: W.W. Norton.

Gefen, D., Benbasat, I., & Pavlou, P. A. (2008). A research agenda for trust in online environments. *Journal of Management Information Systems, 24*(4), 275–286 (Spring).

Gefen, D., Karahanna, E., & Straub, D. W. (2003, March). Trust and TAM in online shopping: An integrated model. *MIS Quarterly, 27*(1), 51–90.

Ghose, A., Ipeirotis, P. G., & Sundararajan, A. (2005, August). Reputation premiums in electronic peer-to-peer markets: Analyzing textual feedback and network structure. In *Proceedings of the 2005 ACM SIGCOMM Workshop on Economics of Peer-to-Peer Systems, Philadelphia, PA*, 22–26 (pp. 150–154).

Grill-Spector, K., Henson, R., & Martin, A. (2006). Repetition and the brain: neural models of stimulus-specific effects. *Trends in Cognitive Sciences, 10*, 14–23.

Henson, R. N., & Rugg, M. D. (2003). Neural response suppression, haemodynamic repetition effects, and behavioural priming. *Neuropsychologia, 41*, 263–270.

Holzwarth, M., Janiszewski, C., & Neumann, M. M. (2006, October). The influence of avatars on online consumer shopping behavior. *Journal of Marketing, 70*(4), 19–36.

Huang, Y., & Rao, R. P. N. (2011). Predictive coding. *Wiley Interdisciplinary Reviews: Cognitive Science, 2*(5), 580–593.

Ishii, A., Tanaka, M., Shigihara, Y., Kanai, E., Funakura, M., & Watanabe, Y. (2013). Neural effects of pronolnged mental fatigue: A magnetoencephalography study. *Brain Research, 1529*, 105–112.

Kilner, J. M., Friston, K. J., & Frith, C. D. (2007). The mirror-neuron system: A Bayesian perspective. *NeuroReport, 18*(6), 619–623.

King, M. G., Burrows, G. D., & Stanley, G. V. (1983). Measurement of stress and arousal: Validation of the stress arousal adjective checklist. *British Journal of Psychology, 74*, 473–479.

Kock, N. (2004, May–June). The psychological model: Towards a new theory of computer-mediated communication based on Darwinian evolution. *Organization Science, 15*(3), 327–348.

Kock, N. (2005). Media richness or media naturalness? The evolution of our biological communication apparatus and its influence on our behavior toward e-communication tools. *IEEE Transactions on Professional Communication, 48*(2), 117–130.

Kock, N. (2009, June). Information systems theorizing based on evolutionary psychology: An interdisciplinary review and theory integration framework. *MIS Quarterly, 33*(2), 395–418.

Kolb, B., & Wishaw, I. Q. (2009). *Fundamentals of Human Neuropsychology* (5th ed.). Palgrave Macmillan.

Ku, J., Jang, H. J., Kim, K. U., Kim, J. H., Park, S. H., & Lee, J. H., et al. (2005, October). Experimental results of affective valence and arousal to avatar's facial expressions. *CyberPsychology and Behavior, 8*(5), 493–503.

Lewicki, R. J., McAllister, D. J., & Bies, R. J. (1998). Trust and distrust: New relationships and realities. *Academy of Management Review, 23*(3), 438–458.

Lewis, P. A., Critchley, H. D., Rotshtein, P., & Dolan, R. J. (2007). Neural correlates of processing valence and arousal in affective words. *Cerebral Cortex, 17*(3), 742–748.

Lewis, J. D., & Weigert, A. (1985, June). Trust as a social reality. *Social Forces, 63*(4), 967–985.

Mackay, C., Cox, T., Burrows, G., & Lazzerini, T. (1978). Inventory for measurement of self-reported stress and arousal. *British Journal of Social and Clinical Psychology, 17*, 283–284.

McKnight, D. H., & Chervany, N. L. (2001). Trust and distrust definitions: One bite at a time. In R. Falcone, M. Singh, & Y.-H. Tan (Eds.), *Trust in cyber-societies* (pp. 27–54), Springer, Berlin.

McKnight, H. D., Cummings, L. L., & Chervany, N. L. (1998, July). Initial trust formation in new organizational relationships. *The Academy of Management Review, 23*(3), 473–490.

Moody, G. D., Galletta, D. F., & Lowry, P. B. (2014). When trust and distrust collide online: The engenderment and role of consumer ambivalence in online consumer behavior. *Electronic Commerce Research and Applications, 13*, 266–282.

Morris, J. et al. (1996). A differential neural response in the human amygdala to fearful and happy facial expressions. *Nature, 383*, 812–815.

Moser, E., Derntl, B., Robinson, S., Fink, B., Gur, R. C., & Grammer, K. (2007, March). Amygdala activation at 3T in response to human and avatar facial expressions of emotions. *Journal of Neuroscience Methods, 161*(1), 126–133.

Nowak, K. L., & Rauh, C. (2005, November). The influence of the avatar on online perceptions of anthropomorphism, androgyny, credibility, homophily, and attraction. *Journal of Computer-Mediated Communication, 11*(1), 153–178.

Pavlou, P. A., & Dimoka, A. (2006). The nature and role of feedback text comments in online marketplaces: Implications for trust building, price premiums, and seller differentiation. *Information Systems Research, 17*(4), 392–414.

Pavlou, P. A., Liang, H., & Xue, Y. (2007, March). Understanding and mitigating uncertainty in online exchange relationships: A principal-agent perspective. *MIS Quarterly, 31*(1), 105–136.

Phelps, E. A., & LaBar, K. S. (2006). Functional neuroimaging of emotion and social cognition. In R. Cabeza & A. Kingstone (Eds.), *Handbook of Functional Neuroimaging of Cognition* (2nd ed., pp. 421–453). Cambridge, MA: MIT Press.

Ponz, A., Montant, M., Liegeois-Chauvel, C., Silva, C., Braun, M., Jacobs, A. M., et al. (2014). Emotion processing in words: A Test of the neural re-use hypothesis using surface and intracranial EEG. *Social Cognitive and Affective Neuroscience, 9*, 619–627.

Qiu, L., & Benbasat, I. (2005, January). Online consumer trust and live help interfaces: The effects of text-to-speech voice and three-dimensional avatars. *International Journal of Human-Computer Interaction, 19*(1), 75–94.

Qiu, L., & Benbasat, I. (2009). Evaluating anthropomorphic product recommendation agents: A social relationship perspective to designing information systems. *Journal of Management Information Systems, 25*(4), pp. 145–182. Spring.

Rao, R. P. N., & Ballard, D. H. (1999). Predictive coding in the visual cortex: A functional interpretation of some extra-classical receptive-field effects. *Nature Neuroscience, 2*(1), 79–87.

Riedl, R., Hubert, M., & Kenning, P. (2010, June). Are there neural gender differences in online trust? An fMRI study on the perceived trustworthiness of eBay offers. *MIS Quarterly, 34*(2), 397–428.

Riegelsberger, J., Sasse, A. M., & McCarthy, J. D. (2003). Shiny happy people building trust? Photos on e-commerce websites and consumer trust. In *Proceedings of CHI 2003, Ft. Lauderdale, Florida*, April 5–10 (pp. 121–128).

Russell, J. A. (2003). Core affect and the psychological construction of emotion. *Psychological Review, 110*(1), 145–172.

Saygin, A. P., Chaminade, T., Ishiguro, H., Driver, J., & Frith, C. (2012). The thing that should not be: predictive coding and the uncanny valley in perceiving human and humanoid robot actions. *Social Cognitive and Affective Neuroscience, 7*(4), 413–422.

Schul, Y., Mayo, R., & Burnstein, E. (2004, May). Encoding under trust and distrust: The spontaneous activation of incongruent cognitions. *Journal of Personality and Social Psychology, 86*(5), 668–679.

Schul, Y., Mayo, R., & Burnstein, E. (2008, September). The value of distrust. *Journal of Experimental Social Psychology, 44*(5), 1293–1302.

Schultz, W. (2006, January). Behavioral theories and the neurophysiology of reward. *Annual Review of Psychology, 57*, 87–115.

Secord, P. F. (1958). Facial features and inference processes in interpersonal perception. In R. Tagiuri & L. Petrullo (Eds.), *Person perception and interpersonal pehavior*. Stanford, CA: Stanford University Press.

Trimmel, M., Meixner-Pendleton, M., & Haring, S. (2003). Stress response caused by system response time when searching for information on the Internet. *Human Factors, 45*(4), 615–621.

Vicente, K. J., Thornton, D. C., & Moray, N. (1987). Spectral analysis of sinus arrhythmia: A measure of mental effort. *Human Factors, 29*(2), 171–182.

Weibel, D., Stricker, D., Wissmath, B., & Mast, F. W. (2010). How socially relevant visual characteristics of avatars influence impression formation. *Journal of Media Psychology, 22*(1), 37–43.

Weyers, P., Mühlberger, A., Hefele, C., & Pauli, P. (2006, September). Electromyographic responses to static and dynamic avatar emotional facial expressions. *Psychophysiology, 43*(5), 450–453.

Winston, J. S., Strange, B. A., O'Doherty, J., & Dolan, R. J. (2002, March). Automatic and intentional brain responses during evaluation of trustworthiness of faces. *Nature Neuroscience, 5*(3), 277–283.

Yee, N., Bailenson, J. N., & Rickertsen, K. (2007). A meta-analysis of the impact of the inclusion and realism of human-like faces on user experiences in interfaces. In *CHI '07 Proceedings of the Conference on Human Factors in Computing Systems, San Jose, California*, April 28–May 3 (pp. 1–10).

Zacharia, G., & Maes, P. (2000, October). Trust management through reputation mechanisms. *Applied Artificial Intelligence Journal, 14*(9), 881–908.

Zucker, L. G. (1986). Production of trust: Institutional sources of economic structure. In B. M. Staw & L. L. Cummings (Eds.), *Research in organizational behavior* (8, pp. 53–111). Greenwich, CT: JAI Press.

Chapter 4
Notes on the Application of the Approach

Application of our approach implies identification, processing, and use of neuroscience knowledge. In particular, the IS researcher must acquire knowledge on the neural correlates of the constructs of his or her study.[1] But where can IS researchers find such knowledge? As established in Riedl et al. (2010a, p. 254), this kind of neuroscience knowledge is often published in classic neuroscience journals, making it essential for IS scholars to know the mainstream neuroscience outlets (Riedl et al. list 15 journals, such as *Cerebral Cortex, Human Brain Mapping, Journal of Neuroscience, NeuroImage,* or *Neuron* as example sources for the IS community). However, reading the neuroscience literature can be demanding—particularly so because the number of neuroscience journals is extensive. Accordingly, the quest to identify the desired neural correlates of a specific IS construct can be a time-consuming process.[2]

---

[1]Note that we use the term "neural correlate" to refer to research based on tests of association, necessity, and sufficiency (see Fig. 2.1). Thereby, we seek to avoid complex sentence structures because, in the case of a test of necessity, a "neural correlate" turns to a "neural determinant" of a mental process. Moreover, we recommend a paper by Camerer and Smith (2013) that discusses the question of whether cognitive functions are localizable in the brain.

[2]A journal search via SCImago Journal and Country Rank yielded 96 hits (search terms: Journal Title = "Neuroscience"; date = July 13, 2015). Considering that many neuroscience journals do not have the term "neuroscience" in the journal name (see the five examples in the main text, where only "Journal of Neuroscience" would meet this criterion), it can be assumed that several hundred relevant journals exist.

© Springer International Publishing AG 2017
R. Riedl et al., *Neuroscience in Information Systems Research*,
Lecture Notes in Information Systems and Organisation 21,
DOI 10.1007/978-3-319-48755-7_4

Moreover, the terms and abstraction levels used to describe a specific phenomenon may vary, and some disciplines (as a function of their research foci) more carefully distinguish related concepts than do other disciplines (neuroeconomics, for example, sharply distinguishes between the concepts of risk, ambiguity, and uncertainty).[3] From an IS perspective, it follows that plain keyword searches via journal databases may not always yield high-quality results, primarily because construct names, or definitions of constructs, may vary across discipline boundaries. For example, an IS researcher interested in uncertainty in online environments is likely to overlook relevant brain research studies if the searches are limited to the neural correlates of the term "uncertainty." Thus, we recommend that as a default position, IS scholars extend keyword searches to related concepts and synonyms, thereby increasing the probability of identifying relevant cognitive neuroscience papers. For example, IS researchers interested in "uncertainty" should also search for "risk" and "ambiguity," those interested in "anxiety" should also search for "phobia" and "fear," and those interested in pleasure should also search for "happiness," "enjoyment," "utility," and "reward".

Over the past few years, brain researchers, cognitive scientists, and psychologists have started with the development of cognitive ontologies and cognitive atlases, respectively. The general objective of these projects is insightfully summarized in a description by Poldrack et al. (2011, p. 1) of the *Cognitive Atlas* project (www. cognitiveatlas.org): "Cognitive neuroscience aims to map mental processes onto brain function, which begs the question of what 'mental processes' exist and how they relate to the tasks that are used to manipulate and measure them. This topic has

---

[3]Uncertainty is used as the more abstract concept, and includes risk (uncertainty with known probabilities) and ambiguity (uncertainty with known probabilities) (e.g., Huettel et al. 2006). With respect to differences in abstraction levels, consider the following example: To investigate a certain mental process (MP_1), a researcher conducts a brain imaging study based on a certain stimulus (referred to as S_1) and finds significantly more activity in a certain brain region (BR_1), as compared to a control condition. In other words, MP_1 is correlated with BR_1. In another study, activity is found in BR_1 as well, but this study investigates a different mental process, namely MP_2. Finally, a stream of research finds activation in BR_1 as a neural response to MP_3. Based on the principle of abstraction, it could be theorized that MP_1, MP_2, and MP_3 are instances of a more abstract mental process. For example, if (i) MP_1 was unfairness (Sanfey et al. 2003), (ii) MP_2 was distrust (Winston et al. 2002), and (iii) MP_3 was risk (Preuschoff et al. 2008), and if BR_1 was the insular cortex, then it could be theorized that the abstract mental process is that of a negative emotional state. A cognitive neuroscience hypothesis would thus state: The insular cortex is a brain region that encodes negative emotional states in humans. Using the analogy of the function of a knife blade, Poldrack (2010, p. 756) illustrates the importance of abstraction in cognitive neuroscience research: "[I]magine a group of people individually trying to understand the function of a knife blade. One person tests its ability to cut peaches. Upon finding that the blade cuts through peach flesh but not through the pit, he or she concludes that the knife is specialized for peach flesh removal. Another person might test its ability to screw various types of screws; finding that the knife blade works well to screw flathead and Phillips screws but not Allen screws, he or she might conclude that it is specialized for a subset of screwing functions. Although each of these is a valid description of the functions that the knife participates in, neither seems to be an accurate description of the fundamental function of a knife blade, such as 'cutting or manipulating objects depending upon their hardness'".

been addressed informally in prior work, but we propose that cumulative progress in cognitive neuroscience requires a more systematic approach to representing the mental entities that are being mapped to brain function and the tasks used to manipulate and measure mental processes." The phrase "more systematic approach" primarily means that online platforms are provided for scholars to contribute their research findings in a structured way.

The main benefit of ontologies and atlases such as are found at the *Cognitive Atlas* project is that researchers and, accordingly, IS scholars as well, can systematically search for specific mental processes (concepts), tasks, and related brain systems. Moreover, some of the platforms (e.g., *Cognitive Atlas*) also provide information on ontological relations (e.g., *is-a* [for example, social phobia is a kind of fear], *part-of* [spatial attention is a part of attention], or *preceded-by* [e.g., memory consolidation is preceded by memory encoding]).[4] Obviously, this kind of cognitive neuroscience knowledge documentation provides a valuable basis for IS scholars who are in search of cognitive science concepts (many of which constitute, or resemble, IS constructs) and corresponding brain systems. In addition to the mentioned example of *Cognitive Atlas*, other example platforms are *Cognitive Paradigm Ontology* (www.cogpo.org), *BrainMap* (www.brainmap.org), and *Neurosynth* (www.neurosynth.org).

With respect to the Qiu and Benbasat (2005) study, as compared to the Cyr et al. (2009) and Pavlou and Dimoka (2006) studies, a notable difference must be highlighted. Predictive coding (or prediction error) is a different form of neuroscience knowledge if compared to knowledge on specific neural correlates of a mental process (i.e., "activity in brain area N is related to …"). It follows that IS researchers should not only concentrate on studying the specific functional roles of different brain areas (as discussed in the previous paragraphs), but they should also acquire neuroscience knowledge about more general brain properties, such as predictive coding. We have indicated that knowledge on neural correlates of mental processes that constitute (or resemble) IS constructs is well documented in cognitive ontologies and atlases. However, the IS researcher should also know where to find knowledge on more fundamental brain properties.

To the best of our knowledge, systematic documentation of fundamental brain properties similar to cognitive ontologies and atlases do not exist. Thus, IS researchers need to conduct literature reviews using specific keywords and database searches (keywords should be derived from the research question at hand). Searching the indexes of seminal textbooks (e.g., Cacioppo et al. 2007 or Gazzaniga et al. 2013) is a recommended complementary strategy.

For example, Riedl et al. (2010b) studied gender differences in online trust and used knowledge of gender differences in (i) brain anatomy (e.g., the female hippocampus is, on average, larger than that of the male; Cosgrove et al. 2007) and (ii) brain functioning (e.g., empathizing–systemizing framework; Baron-Cohen

---

[4]Examples taken from http://www.cognitiveatlas.org/ (July 13, 2015).

et al. 2005) for their theorizing. This neuroscience knowledge was identified through use of keyword searches in journal databases.[5] Importantly, while the Riedl et al. (2010b) study on gender differences in brain anatomy and brain functioning uses fMRI, the neuroscience knowledge discussed in their paper can also be used to develop hypotheses that can be tested on a solely behavioral level—a fact of particular importance to IS scholars who do not have access to MRI machines but who still want to benefit from cognitive neuroscience. For example, it is an anatomical fact that the female hippocampus is, on average, larger than that of the male (Cosgrove et al. 2007), and because the hippocampus is closely related to emotional memory (e.g., Gazzaniga et al. 2009, pp. 374–387), this anatomical gender difference is likely to contribute to women's heightened perception of risk (because negative experiences are better remembered by women) (Byrnes et al. 1999). Consistent with our framework in Fig. 3.1, this example gender difference in risk perception can be tested on a solely behavioral level (based on behavioral paradigms or rating tasks).

In addition to this occasion-driven strategy (i.e., a search for general brain properties based on a specific IS research question), we recommend that IS researchers interested in NeuroIS read neuroscience literature systematically, giving particular attention to the latest editions of seminal textbooks in the fields of cognitive neuroscience (e.g., Gazzaniga et al. 2013), neuropsychology (e.g., Kolb and Whishaw 2015), and psychophysiology (e.g., Cacioppo et al. 2007). These books can be supposed to document hundreds of general brain properties, ranging from anatomical knowledge (e.g., brain area X has reciprocal connections with brain area Y; brain area Z, on average, is anatomically larger in women than in men; or brain area Z declines with age) to highly specific theories such as the somatic marker hypothesis, a neural theory of economic decision making (Bechara and Damasio 2005).[6] Building on the basis of reading about or investigating a wide range of brain properties and related cognitive neuroscience knowledge, IS scholars may develop ideas about how that knowledge might shed light on a specific IS phenomenon. This *creative act* of neuroscience knowledge application, however, is different from the described strategy wherein the IS research question precedes neuroscience knowledge acquisition (for example, through keyword-based database searches). Here, the IS researcher first becomes familiar with fundamental brain properties and cognitive neuroscience knowledge (for example, through reading

---

[5]Note that these two types of knowledge are examples for what we refer to as general brain properties in the present book; in other words, the body of neuroscience knowledge except knowledge on the specific functions of brain areas.

[6]The somatic marker hypothesis (SMH) is explicitly mentioned in Dimoka et al.'s (2012) NeuroIS research agenda paper as a relevant theory from the cognitive neuroscience literature (see Appendix B). Moreover, Léger et al. (2014) use the SMH as a conceptual basis in a NeuroIS study on emotions exhibited during use of an enterprise resource planning (ERP) system. Also, it is important to mention that the development of the SMH hypothesis has been based on tests of association and necessity (Bechara and Damasio 2005).

seminal textbooks with no specific research context in mind). Through consequent creative thought processes, researchers are able to identify IS context areas where the neuroscience knowledge can be applied.

References

Baron-Cohen, S., Knickmeyer, R. C., & Belmonte, M. K. (2005). Sex differences in the brain: Implications for explaining Autism. *Science, 310*(5749), 819–823.

Bechara, A., & Damasio, A. R. (2005, May). The somatic marker hypothesis: A neural theory of economic decision. *Games and Economic Behavior, 52*(2), 336–372.

Byrnes, J. P., Miller, D. C., & Schafer, W. D. (1999). Gender differences in risk taking: A meta-analysis. *Psychological Bulletin, 125*(3), 367–383.

Cacioppo, J. T., Tassinary, L. G., & Berntson, G. G. (2007). *Handbook of psychophysiology* (3rd ed.). Cambridge: Cambridge University Press.

Camerer, C., & Smith, A. (2013). Are cognitive functions localizable? *Journal of Economic Perspectives, 27*(2), 247–252.

Cosgrove, K. P., Mazure, C. M., & Staley, J. K. (2007). Evolving knowledge of sex differences in brain structure, function, and chemistry. *Biological Psychiatry, 62*(8), 847–855.

Cyr, D., Head, M., Larios, H., & Pan, B. (2009, September). Exploring human images in website design: A multi-method approach. *MIS Quarterly, 33*(3), 539–566.

Dimoka, A., Banker, R. D., Benbasat, I., Davis, F. D., Dennis, A. R., Gefen, D., et al. (2012). On the use of neurophysiological tools in IS research: Developing a research agenda for NeuroIS. *MIS Quarterly, 36*(3), 679–702.

Gazzaniga, M. S., Ivry, R., & Mangun, G. R. (2009). *Cognitive neuroscience: The biology of the mind* (3rd ed.). New York: W.W. Norton.

Gazzaniga, M. S., Ivry, R., & Mangun, G. R. (2013). *Cognitive neuroscience: The biology of the mind* (4th ed.). New York: W.W. Norton.

Huettel, S. A., Stowe, C. J., Gordon, E. M., Warner, B. T., & Platt, M. L. (2006, March). Neural signatures of economic preferences for risk and ambiguity. *Neuron, 49*(5), 765–775.

Kolb, B., & Wishaw, I. Q. (2015). *Fundamentals of human neuropsychology* (7th ed.). New York: Worth Publishers.

Léger, P.-M., Riedl, R., & vom Brocke, J. (2014). Emotions and ERP information sourcing: The moderating role of expertise. *Industrial Management and Data Systems, 114*(3), 456–471.

Pavlou, P. A., & Dimoka, A. (2006, December). The nature and role of feedback text comments in online marketplaces: Implications for trust building, price premiums, and seller differentiation. *Information Systems Research, 17*(4), 392–414.

Poldrack, R. A. (2010, November). Mapping mental function to brain structure: How can cognitive neuroimaging succeed? *Perspectives on Psychological Science, 5*(6), 753–761.

Poldrack, R. A., Kittur, A., Kalar, D., Miller, E., Seppa, C., Gil, Y., et al. (2011). The cognitive atlas: Toward a knowledge foundation for cognitive neuroscience. *Frontiers in Neuroinformatics, 5*(17), 1–11.

Preuschoff, K., Quartz, S. R., & Bossaerts, P. (2008, March). Human insula activation reflects risk prediction errors as well as risk. *Journal of Neuroscience, 28*(11), 2745–2752.

Qiu, L., & Benbasat, I. (2005, January). Online consumer trust and live help interfaces: The effects of text-to-speech voice and three-dimensional avatars. *International Journal of Human-Computer Interaction, 19*(1), 75–94.

Riedl, R., Banker, R. D., Benbasat, I., Davis, F. D., Dennis, A. R., Dimoka, A., et al. (2010a). On the foundations of NeuroIS: Reflections on the Gmunden Retreat 2009. *Communications of the Association for Information Systems, 27*(15), 243–264.

Riedl, R., Hubert, M., & Kenning, P. (2010b, June). Are there neural gender differences in online trust? An fMRI study on the perceived trustworthiness of eBay offers. *MIS Quarterly, 34*(2), 397–428.

Sanfey, A. G., Rilling, J. K., Aronson, J. A., Nystrom, L. E., & Cohen, J. D. (2003, June). The neural basis of economic decision-making in the ultimatum game. *Science, 300*(5626), 1755–1758.

Winston, J. S., Strange, B. A., O'Doherty, J., & Dolan, R. J. (2002, March). Automatic and intentional brain responses during evaluation of trustworthiness of faces. *Nature Neuroscience, 5*(3), 277–283.

Chapter 5
Conclusion

The field of *NeuroIS* has made advancements during the recent past. Conceptual articles (e.g., Dimoka et al. 2011, 2012; Loos et al. 2010; Riedl 2009; Riedl et al. 2010), empirical studies (see Appendix A), and guideline papers (Dimoka 2012; Hubert et al. 2017; Müller-Putz et al. 2015; Riedl et al. 2014; vom Brocke and Liang 2014) have contributed to a better understanding of the relationship between brain processes and IS behavior. Importantly, the establishment of NeuroIS is associated with both the use of neuroscience tools and the application of existing cognitive neuroscience knowledge.

In this book, we have focused on cognitive neuroscience knowledge. We briefly outlined major knowledge production processes in cognitive neuroscience in order to develop a conceptual basis for the sections to follow. Based on three concrete IS research examples in the domain of online trust, we described how IS scholars can apply knowledge on brain functions without necessarily using neuroscience tools in order to better understand IS phenomena. Drawing on these three examples, we formalized our approach so that it can be used for future research by other IS scholars, independent of their specific area of study. Moreover, we outlined recommendations related to the application of the approach presented in this paper. Altogether, the present book demonstrates that cognitive neuroscience knowledge makes a more comprehensive understanding of IS phenomena possible by connecting the behavioral and brain levels of analysis, thereby facilitating advancements in the IS discipline.

The employment of advanced neuroscience tools (e.g., fMRI, TMS, lesion studies) in disciplines such as cognitive neuroscience has resulted in the production of a vast amount of knowledge exploring the brain mechanisms underlying human behavior. We argue that it is possible for IS scholars to benefit from cognitive neuroscience literature without using neuroscience tools. In a pioneering article of this field, Dimoka et al. (2007, p. 1) wrote that "IS researchers can become intelligent users of this emerging [cognitive neuroscience] literature to strengthen the theorizing". Our book presents examples of ways in which the behavioral and brain levels of analysis can be connected in order to gain a deeper understanding of IS

© Springer International Publishing AG 2017
R. Riedl et al., *Neuroscience in Information Systems Research*,
Lecture Notes in Information Systems and Organisation 21,
DOI 10.1007/978-3-319-48755-7_5

phenomena, and formalizes the logic of our argument so that other IS scholars can apply this approach.

We close our analysis of the efficacy and importance of cognitive neuroscience knowledge for strengthening IS research with a look back to the history of scientific exploration in mature disciplines such as astronomy and biology (Camerer et al. 2005). In the late 16th century, Dutch spectacle makers discovered precursor technologies of what we now call the telescope and microscope. These technologies were further developed by authorities such as Galileo Galilei (1564–1642), yet only a very limited number of individuals in that era had the opportunity to use the technologies. Despite the fact that most people never had the opportunity to observe objects via these new technologies, reports from the experience of others—whether conveyed orally or in written form—elevated society's understanding of the world. In this way, the understandings that people had developed about the physical world through simple visual perception, without the benefit of technologies, were enhanced by the information gathered and communicated by those who, through the technologies of the time, had access to another level of analysis (e.g., Wolfschmidt 2004). Importantly, even though several centuries have passed since technological investigations enhanced knowledge in this way, the conditions demonstrated by that research support the similarity of present-day research conditions. Not every IS researcher, for example, has access to an fMRI scanner, or to other state-of-the-art brain science technologies such as TMS. Yet the vast amount of cognitive neuroscience literature (i.e., reports from other scholars who have used the technology for research) makes learning about a new level of analysis possible. Connecting these novel insights with existing (behavioral) knowledge is likely to result in important new body of knowledge—in the confirmation, rejection, or refinement of existing hypotheses and theories. It is difficult to imagine that IS theorizing would not be enhanced by a new level of observation such as that contributed by cognitive neuroscience studies. Whether the brain sciences will change IS theorizing as much as the telescope and microscope changed theories in astronomy and biology cannot be predicted, but the change itself is certain.

Acknowledgments We thank all scholars from the IS and neuroscience disciplines, as well as six anonymous reviewers, for their work in providing guidance on ways to improve the paper. Also, we thank Deborah C. Nester for proof-reading.

References

Camerer, C., Loewenstein, G., & Prelec, D. (2005, March). Neuroeconomics: How neuroscience can inform economics. *Journal of Economic Literature, 43*(1), 9–64.

Dimoka, A. (2012). How to conduct a functional magnetic resonance (fMRI) study in social science research. *MIS Quarterly, 36*(3), 811–840.

Dimoka, A., Banker, R. D., Benbasat, I., Davis, F. D., Dennis, A. R., Gefen, D., et al. (2012). On the use of neurophysiological tools in IS research: Developing a research agenda for NeuroIS. *MIS Quarterly, 36*(3), 679–702.

Dimoka, A., Pavlou, P. A., & Davis, F. D. (2007). NEURO-IS: The potential of cognitive neuroscience for information systems research. In *ICIS 2007*.

Dimoka, A., Pavlou, P. A., & Davis, F. D. (2011, December). NeuroIS: The potential of cognitive neuroscience for information systems research. *Information Systems Research, 22*(4), 687–702.

Hubert, M., Linzmajer, M., Riedl, R., Kenning, P., & Weber, B. (2017). The use of psycho-physiological interaction analysis with fMRI-data in IS research: A guideline. *Communications of the Association for Information Systems*, forthcoming 2017.

Loos, P., Riedl, R., Müller-Putz, G. R., vom Brocke, J., Davis, F. D., Banker, R., et al. (2010, December) "NeuroIS: Neuroscientific approaches in the investigation and development of information systems. *Business and Information Systems Engineering, 2*(6), 395–401.

Müller-Putz, G. R.; Riedl, R., & Wriessnegger, S. C. (2015). Electroencephalography (EEG) as a research tool in the information systems discipline: Foundations, measurement, and applications. *Communications of the Association for Information Systems, 37*, 911–948.

Riedl, R. (2009). Zum Erkenntnispotenzial der kognitiven Neurowissenschaften für die Wirtschaftsinformatik: Überlegungen anhand exemplarischer Anwendungen. *NeuroPsycho Economics* (4), 32–44.

Riedl, R., Banker, R. D., Benbasat, I., Davis, F. D., Dennis, A. R., Dimoka, A., et al. (2010). On the foundations of NeuroIS: Reflections on the Gmunden Retreat 2009. *Communications of the Association for Information Systems, 27*(15), 243–264.

Riedl, R., Davis, F. D., & Hevner, A. R. (2014). Towards a NeuroIS research methodology: Intensifying the discussion on methods, tools, and measurement. *Journal of the Association for Information Systems, 15*(10) (Article 4).

vom Brocke, J., & Liang, T. P. (2014). Guidelines for neuroscience studies in information systems research. *Journal of Management Information Systems, 30*(4), 211–234.

Wolfschmidt, G. (2004). Die Eroberung des Himmels. In R. Van Dülmen & S. Rauschenbach (Eds.), *Macht des Wissens: Die Entstehung der modernen Wissensgesellschaft* (pp. 187–212). Wien: Böhlau.

Part II
Appendix

Appendix A: Review of Empirical NeuroIS Literature

To identify empirical NeuroIS papers published in peer-reviewed journals, we searched for articles via WEB OF KNOWLEDGE^{SM} (June 30, 2015). Specifically, we searched for "NeuroIS" based on the constraint <TOPIC>. The search resulted in 20 hits. Seven articles were excluded from further analysis because they are not empirical contributions (i.e., research articles): Loos et al. (2010) (discussion paper), Dimoka et al. (2011) (research commentary), Liapis and Chatterjee (2011) (conceptual paper), Dimoka (2012) (research essay), Dimoka et al. (2012) (issues and opinions paper), vom Brocke and Liang (2014) (guideline paper), and Riedl et al. (2014) (special issue editorial).

Moreover, we sought to identify further empirical NeuroIS papers published in peer-reviewed IS journals. To this end, we searched for such contributions in the eight IS outlets belonging to the "Senior Scholars' Basket of Journals" (for details, see http://aisnet.org/?SeniorScholarBasket). We started the search in 2008, following the emergence of the concept of NeuroIS in December 2007 (Dimoka et al. 2007); the search ended on June 30, 2015. Based on this additional search, we identified two further papers: Ortiz de Guinea and Webster (2013) and Riedl et al. (2010).

The order of the resulting 15 papers in the Table A.1 is based on date of publication. In the column "Construct(s)," we indicate the constructs which were physiologically measured (note that this does not rule out the possibility that a construct was also measured by other methods, such as a survey). All studies used pre- and/or post-experiment questionnaires (e.g., to collect demographic data). However, only in the cases in which a questionnaire was used to measure a construct (before, during, or after physiological measurement) that was also part of the statistical analysis (as independent, dependent, mediator, or moderator variable) do we indicate "survey" in the "Measurement Instrument(s)" column. Here, then, "survey" means a self-reported perception of a construct. Some articles report on more than one study. The indication "study not directly relevant in the present context" means that a specific study does not include physiological measurement, and hence is not described in detail in the Table A.1.

© Springer International Publishing AG 2017
R. Riedl et al., *Neuroscience in Information Systems Research*,
Lecture Notes in Information Systems and Organisation 21,
DOI 10.1007/978-3-319-48755-7_6

A general conclusion of the analysis of the extant literature is that empirical NeuroIS research has investigated many different constructs. Thus, the thematic diversity in contemporary NeuroIS research is high. Table A.1 gives a summary of our review of empirical NeuroIS literature. Future review studies should use our analysis as a starting point for updates, which include NeuroIS papers published after June 30, 2015.

Table A.1 Summary of review of empirical NeuroIS literature

Study 1	Topic	Construct(s)	Research method
Dimoka (2010)	Trust in online environments	Trust (credibility, benevolence) Distrust (discredibility, malevolence)	Laboratory experiment (Within-subject)
Measurement instrument(s)	**Analysis technique(s)**	**Task**	**Sample and miscellaneous**
fMRI (BOLD signal) Survey	Correlation Regression	Simulated purchase of an MP3 player (eBay)	N = 15 (9m/6f), right-handed (N = 177 in a behavioral study before the main fMRI experiment)

Major result (related to physiology): "[T]rust is associated with the brain's reward [caudate nucleus, putamen], prediction [anterior paracingulate cortex], and uncertainty areas [orbitofrontal cortex], while distrust is associated with the brain's intense emotions [amygdala] and fear of loss areas [insular cortex] … credibility and discredibility are mostly associated with the brain's cognitive areas (prefrontal cortex), while benevolence and malevolence are mostly associated with the emotional areas (limbic system) … the identified brain areas adequately predict price premiums, and the levels of brain activation have a stronger predictive power than the corresponding self-reported psychometric measures" (p. 388)

Study 2	Topic	Construct(s)	Research method
Riedl et al. (2010)	Trust in online environments	Trust (with focus on gender differences)	Laboratory experiment (Within-subject)
Measurement instrument(s)	**Analysis technique(s)**	**Task**	**Sample and miscellaneous**
fMRI (BOLD signal)	Correlation ANOVA	Simulated purchase of an USB flash drive (eBay)	N = 20 (10m/10f), right-handed (N = 39 in a behavioral study before the main fMRI experiment)

Major result (related to physiology): "[W]e found some similarities [e.g., insular cortex related to disgust, uncertainty, or anticipation of pain] and substantial differences [e.g., caudate nucleus, putamen, and thalamus related to reward, or prefrontal structures related to anticipation of future consequences, or hippocampus related to memory] between neural processing in women and men" (p. 415)

Study 3	Topic	Construct(s)	Research method
Nunamaker et al. (2011)	Development and evaluation of an automated kiosk that uses embodied intelligent agents to interview individuals and detect changes in arousal, behavior, and cognitive effort by using psychophysiological information systems	Stress (arousal) (deception, emotion, cognitive effort)	Laboratory experiment (Within-subject)

(continued)

Table A.1 (continued)

Measurement instrument(s)	Analysis technique(s)	Task	Sample and miscellaneous
Vocal measurement (pitch) Survey	Regression ANCOVA	Simulated airport screening interaction (i.e., interview process with the task to answer some questions truthfully and some deceptively)	N = 110 (66m, 44f) (aggregated data of Study 2 and Study 3 reported) Note: Study 1 is not directly relevant in the present context. Hence, only Study 2 and Study 3 are reported here

Major result (related to physiology): "[T]his research demonstrates how even a single sensor [measuring vocal pitch, people speak with a higher pitch and with more variation in pitch or fundamental frequency when under increased stress or arousal], properly modeled, can provide … awareness of human emotion or behavior" (p. 42). "We have demonstrated how manipulations to the embodied states change participants' perceptions of the system and how a sensor can detect emotion and arousal" (p. 46)

Study 4	Topic	Construct(s)	Research method
Riedl et al. (2012)	Stress resulting from system breakdown	Technostress	Laboratory experiment (Between-subject)

Measurement instrument(s)	Analysis technique(s)	Task	Sample and miscellaneous
Stress hormone measurement (salivary cortisol)	ANOVA	Searching for products and putting them into the shopping cart in an online shop	N = 20 (20m)

Major result (related to physiology): "[S]ystem breakdown in the form of an error message is an acute stressor which may elicit cortisol elevations as high as in non-HCI [human-computer interaction] stress situations such as public speaking (e.g., Trier Social Stress Test)" (p. 66)

Study 5	Topic	Construct(s)	Research method
Astor et al. (2013)	Development and evaluation of a neuro-adaptive system (biofeedback) in a financial decision-making context	Arousal Emotion regulation	Laboratory experiment (Between-subject)

Measurement instrument(s)	Analysis technique(s)	Task	Sample and miscellaneous
ECG (heart rate) Survey Decision performance	Regression ANOVA	Auction Game (simulated trading of stocks with the goal to earn money)	N = 104 (76m/28f) (aggregated data of the two studies reported)

Major result (related to physiology): "[O]ur study demonstrates how information systems design science research can contribute to improving financial decision making by integrating physiological data into information technology artifacts" (p. 248) … "we designed, implemented, and evaluated a biofeedback-based NeuroIS tool aimed at supporting decision makers with improving their ERCs [emotion regulation capabilities] … The emotional state is assessed by means of unobtrusive ECG [heart rate] measurements" (p. 267/268)

Study 6	Topic	Construct(s)	Research method
Ortiz de Guinea and Webster (2013)	Influence of discrepant IT events on IS use patterns (i.e., the configuration of a user's emotions, cognitions, and behaviors when interacting with a system to accomplish a task)	Physiological arousal	Laboratory experiment (Between-subject)

(continued)

Table A.1 (continued)

Measurement instrument(s)	Analysis technique(s)	Task	Sample and miscellaneous
ECG (heart rate) Survey Protocol analysis Video (behavior analysis) Task performance	ANOVA	Essay writing with a word processing program	N = 103 (gender distribution not reported) Note: Study 1 is not directly relevant in the present context. Hence, only Study 2 is reported here

Major result (related to physiology): "[T]he adjusting IS use pattern is triggered by discrepant IT events and is characterized by negative affect related to the IT being used, lower physiological arousal, computer-related thoughts, and adaptive behaviors aimed at modifying an aspect of the IT" (p. 1182)

Study 7	Topic	Construct(s)	Research method
Gregor et al. (2014)	User emotions during interaction with websites	Positive emotions Negative emotions	Laboratory experiment (Within-subject)

Measurement instrument(s)	Analysis technique(s)	Task	Sample and miscellaneous
EEG (prefrontal activity: F3, F4) Survey	Regression ANOVA	Looking at home pages for different online travel services (subjects assumed that they were planning a trip)	N = 21 (13m, 8f) Note: Study 1 is not directly relevant in the present context. Hence, only Study 2 is reported here

Major result (related to physiology): "[P]ositive and negative emotion-inducing stimuli were related to positive and negative emotions when viewing the Web sites as indicated by both self-reports and EEG data … positive and negative emotions as measured by both self-reports and EEG measures were linked to e-loyalty to some degree … EEG measure has some predictive power for an outcome such as e-loyalty" (p. 14)

Study 8	Topic	Construct(s)	Research method
Minas et al. (2014)	Text-based information processing of team members during a team decision-making process	Confirmation bias (the tendency to search for or interpret information in a way that confirms one's preconceptions) (operationalized via cortical attention, autonomic arousal, and emotional valence)	Laboratory experiment (Within-subject)

Measurement instrument(s)	Analysis technique(s)	Task	Sample and miscellaneous
EEG (event-related spectral perturbation, ERSP, alpha band frequency, frontal-temporal-occipital (FTO) cluster and left/right frontal cluster) Electrodermal activity Facial electromyography (corrugator supercilii muscle group)	ANOVA t-tests	Decision-making task to select student applicants to admit to the university, including a simulated discussion using a text-based discussion tool	N = 28 (mixed-gender sample and a majority was right-handed)

(continued)

Table A.1 (continued)

Major result (related to physiology): "[F]indings show that information that challenges an individual's prediscussion decision preference is processed similarly to irrelevant information, while information that supports an individual's prediscussion decision preference is processed more thoroughly" (p. 50) "Our findings suggest that a primary cause of poor decision making in virtual teams is confirmation bias, rather than information overload" (p. 76) "lack of statistical significance in the corrugator EMG data … EEG data showed that different information triggered different patterns of cognition, while the EDR data showed that different information triggered different emotional responses" (p. 77)

Study 9	Topic	Construct(s)	Research method
Riedl et al. (2014)	Trust in avatars and humans	Trust (with focus on mentalizing) (trustworthiness prediction ability, mentalizing, trustworthiness learning rate)	Laboratory experiment (Within-subject)
Measurement instrument(s)	**Analysis technique(s)**	**Task**	**Sample and miscellaneous**
fMRI (BOLD signal) Survey	Correlation Regression	Trust Game with avatar faces and human faces (multi-round version)	N = 18 (11m/7f), right-handed (N = 45 in a behavioral study before the main fMRI experiment)

Major result (related to physiology): "[R]esults indicate that (1) people are better able to predict the trustworthiness of humans than the trustworthiness of avatars; (2) decision making about whether or not to trust another actor activates the medial frontal cortex significantly more during interaction with humans, if compared to interaction with avatars; this brain area is of paramount importance for the prediction of other individuals' thoughts and intentions (mentalizing), a notably important ability in trust situations; and (3) the trustworthiness learning rate is similar, whether interacting with humans or avatars" (p. 84)

Study 10	Topic	Construct(s)	Research method
Li et al. (2014)	Enhancement of user game engagement through software gaming elements	Engagement (game complexity, game familiarity) Emotions of game playing	Laboratory experiment (Between-subject)
Measurement instrument(s)	**Analysis technique(s)**	**Task**	**Sample and miscellaneous**
EEG (density of theta oscillation from the left side of the dorsolateral prefrontal cortex (DLPFC) Survey	ANCOVA	Computer game playing using a smartphone with a 3.5-inch screen and earphones	N = 44 (21m, 23f), right-handed Note: Study 2 is not directly relevant in the present context. Hence, only Study 1 is reported here

Major result (related to physiology): "[W]e proposed and demonstrated that cognitive-related gaming elements, which are classified as game complexity and game familiarity, influence the density of theta oscillations from the left side of the DLPFC and game engagement" (p. 138) "we observed that cognitive-related gaming elements can be utilized well in different conditions and can jointly influence user–game engagement … we observed that low game complexity and high game familiarity significantly increase the level of user–game engagement. This significant change is measured by comparing the temporal variances of the density of theta oscillations from the left side of the DLPFC" (p. 142) … "[P]laying software games induces positive emotions, as expected, rendering the left frontal cortex" (p. 129)

(continued)

Table A.1 (continued)

Study 11	Topic	Construct(s)	Research method
Kuan et al. (2014)	Informational and normative social influence in group-buying in online environments	Positive emotions Negative emotions	Laboratory experiment (Within-subject)

Measurement instrument (s)	Analysis technique(s)	Task	Sample and miscellaneous
EEG (low alpha activity in the left frontal region for positive emotions, and low alpha activity in the right frontal region for negative emotions)	ANOVA Regression Correlation	Looking at online deals, including information on the number of people who have bought the deal and who like the deal	N = 18 (7m, 11f), handedness of subjects not reported

Major result (related to physiology): "Our findings show that providing "buy" information in the deals is primarily associated with negative emotional responses in terms of increased high-alpha activity (10–12 Hz) in the left frontal region (F3). However, providing "like" information in addition to "buy" information is associated with positive emotional responses in terms of increased high-alpha activity (10–12 Hz) in the right frontal region (F4). This suggests that the normative social influence exerted by "like" information alleviates the negative experience exerted by "buy" information, rendering the negative emotional responses insignificant. At the same time, providing "like" information triggers the need and desire to be liked and approved by peers and induces a positive motivational drive" (p. 171)

Study 12	Topic	Construct(s)	Research method
Oritz de Guinea et al. (2014)	Explicit and implicit antecedents of users' behavioral beliefs regarding perceived usefulness and perceived ease of use	Distraction Memory load	Laboratory experiment (Within-subject)

Measurement instrument (s)	Analysis technique(s)	Task	Sample and miscellaneous
EEG (distraction: index based on absolute and relative power spectra from channels FzPOz and CzPOz of the theta, alpha, and beta frequencies, memory load: frontal theta activity) Survey	Structural equation modeling (partial least squares) Correlation	Instrumental task using a word processing program and a hedonic task using educational gaming software	N = 24 (gender distribution and handedness of subjects not reported)

Major result (related to physiology): "[W]hen engagement is high, neurophysiological distraction does not statistically significantly affect perceived usefulness, whereas when engagement is low, neurophysiological distraction has a negative and significant effect on usefulness. The results also show that when frustration is high, neurophysiological memory load has a negative effect on perceived ease of use, whereas when it is low, neurophysiological memory load has a positive effect on perceived ease of use" (p. 180)

Study 13	Topic	Construct(s)	Research method
Léger et al. (2014)	Reactions of users to e-mail pop-up notifications	Attention Language processing Motor planning process	Laboratory experiment (Within-subject)

(continued)

Table A.1 (continued)

Measurement instrument(s)	Analysis technique(s)	Task	Sample and miscellaneous
Eye-Fixation Related Potential (EFRP) method, which allows to synchronize eye tracking with EEG recordings	t-tests	Reading task on a computer to prepare for a business meeting, and subjects received e-mails that would either contain relevant or irrelevant task-related information	N = 24 (15m, 9f), handedness of subjects not reported Note: This study also validated the EFRP method (benchmark: Event-Related Potential (ERP) method)

Major result (related to physiology): "[O]ur results clearly show a P300 component at the stimulus onset … which indicates the presence of a bottom-up attentional process" (p. 662), "the mean amplitude of Pz was significantly negative over the considered time interval … which indicates the presence of a N400 component at fixation onset … showing that language-related areas of the brain are involved in this time interval" (p. 663), "results reveal a BP component at response onset indicating the presence of a motor planning process" (p. 663)

Study 14	Topic	Construct(s)	Research method
Anderson et al. (2014)	Users' perception of and response to information security risks	Risk perception	Laboratory experiment (Within-subject)

Measurement instrument(s)	Analysis technique(s)	Task	Sample and miscellaneous
EEG (P300) Survey Click behavior Task performance	Regression t-tests ANOVA	Iowa Gambling Task (an instrument to measure risk-taking behaviors by simulating real-life decision-making)	N = 62 (4 m, 16f)

Major result (related to physiology): "[W]e show that participants' EEG P300 amplitudes in response to losses in a risk-taking experimental task strongly predicted security warning disregard in a subsequent and unrelated computing task using participants' own laptop computers. By comparison, self-reported measures of information security risk did not predict security warning disregard. However, after secretly simulating a malware incident on the participants' own laptops, post-test measures of information security risk perception did predict participants' security warning disregard after a security incident. This suggests that self-reported measures of information security risk can significantly predict security behavior when security risks are salient. In contrast, the P300 risk measure is a significant predictor of security behavior both before and after a security incident is imposed, which highlights the robustness of NeuroIS methods in measuring risk perceptions and their value in predicting security behavior" (p. 704)

Study 15	Topic	Construct(s)	Research method
Tams et al. (2014)	Effects of stress on performance on a computer-based task	Technostress	Laboratory experiment (Between-subject)
Measurement instrument(s)	Analysis technique(s)	Task	Sample and miscellaneous
Stress enzyme measurement (salivary alpha-amylase) Survey	Hierarchical regression analysis	Game-like computer-based task (memory game to find matching pairs of symbols in a matrix by flipping computer-generated cards)	N = 64 (43m, 21f)

(continued)

Table A.1 (continued)

Major result (related to physiology): "We demonstrated for the case of technostress that physiological and psychological measures can diverge. This divergence precludes them from constituting alternative forms of measurement, and it suggests that they could be complementary. We then demonstrated complementarity by using the physiological measure [salivary alpha-amylase] to explain additional variance in performance on a computer-based task, variance to which the psychological measure [self-reported stress measure] was blind. In effect, our physiological measure captured aspects of stress of which the subjects were generally not aware. Hence, the value of NeuroIS research lies in its capacity to complement traditional IS methods so that a more complete understanding of IS phenomena can be obtained and more powerful predictive relationships achieved. Overall, our findings indicate that physiological and psychological measures of such IS constructs as technostress may not be interchangeable and that, in fact, both types of measures together can explain higher levels of variance in IS dependent variables than either one can alone. In doing so, our findings shift the debate on the role of NeuroIS in IS research from one of mere measurement accuracy to one of theoretical richness and more complete prediction and explanation of the consequences of such IS phenomena as technostress." (p. 744)

References

Astor, P. J., Adam, M. T. P., Jerčić, P., Schaaff, K., & Weinhardt, C. (2013). Integrating biosignals into information systems: A NeuroIS tool for improving emotion regulation. *Journal of Management Information Systems, 30*(3), 247–278.

Dimoka, A. (2010). What does the brain tell us about trust and distrust? Evidence from a functional neuroimaging study. *MIS Quarterly, 34*(2), 373–396.

Dimoka, A. (2012). How to conduct a functional magnetic resonance (fMRI) study in social science research. *MIS Quarterly, 36*(3), 811–840.

Dimoka, A., Pavlou, P. A., & Davis, F. D. (2007). NEURO-IS: the potential of cognitive neuroscience for information systems research. *ICIS 2007.*

Dimoka, A., Pavlou, P. A., & Davis, F. D. (2011). NeuroIS: the potential of cognitive neuroscience for information systems research. *Information Systems Research, 22*(4), December, 687–702.

Dimoka, A. et al. (2012). On the use of neurophysiological tools in IS research: Developing a research agenda for NeuroIS. *MIS Quarterly, 36*(3), 679–702.

Gregor, S., Lin, A. C. H., Gedeon, T., Riaz, A., & Zhu, D. (2014). Neuroscience and a nomological network for the understanding and assessment of emotions in information systems research. *Journal of Management Information Systems, 30*(4), 13–48.

Kuan, K. K. Y., Zhong, Y., & Chau, P. Y. K. (2014). Informational and normative social influence in group-buying: Evidence from self-reported and EEG data. *Journal of Management Information Systems, 30*(4), 151–178.

Léger, P.-M., Sénecal, S., Courtemanche, F., Ortiz de Guinea, A., Titah, R., Fredette, M., et al. (2014). Precision is in the eye of the beholder: Application of eye fixation-related potentials to information systems research. *Journal of the Association for Information Systems, 15*(10) (Article 3).

Li, M., Jiang, Q., Tan, C.-H., & Wei, K.-K. (2014). Enhancing user-game engagement through software gaming elements. *Journal of Management Information Systems, 30*(4), 115–150.

Liapis, C., & Chatterjee, S. (2011). On a NeuroIS design science model. In H. Jain, A. P. Sinha, & P. Vitharana (Eds.), *Proceedings of the 6th International Conference on Design Science Research in Information Systems and Technology.* DESRIST, LNCS 6629, 440–451.

Loos, P. et al. (2010). NeuroIS: neuroscientific approaches in the investigation and development of information systems. *Business & Information Systems Engineering, 2*(6), December, pp. 395–401.

Minas, R. K., Potter, R. F., Dennis, A. R., Bartelt, V., & Bae, S. (2014). Putting on the thinking cap: Using NeuroIS to understand information processing biases in virtual teams. *Journal of Management Information Systems, 30*(4), 49–82.

Nunamaker, J. F., Derrick, D. C., Elkins, A. C., Burgoon, J. K., & Patton, M. W. (2011). Embodied conversational agent-based kiosk for automated interviewing. *Journal of Management Information Systems, 28*(1), 17–48.

Ortiz de Guinea, A., & Webster, J. (2013). An investigation of information systems use patterns: Technological events as triggers, the effects of time, and consequences for performance. *MIS Quarterly, 37*(4), 1165–1188.

Ortiz de Guinea, A., Titah, R., & Leger, P. M. (2014). Explicit and implicit antecedents of users' behavioral beliefs in information systems: A neuropsychological investigation. *Journal of Management Information Systems, 30*(4), 179–210.

Riedl, R., Hubert, M., & Kenning, P. (2010). Are there neural gender differences in online trust? An fMRI study on the perceived trustworthiness of eBay offers. *MIS Quarterly, 34*(2), 397–428.

Riedl, R., Kindermann, H., Auinger, A., & Javor, A. (2012). Technostress from a neurobiological perspective: System breakdown increases the stress hormone cortisol in computer users. *Business & Information Systems Engineering, 4*(2), 61–69.

Riedl, R., Mohr, P. N. C., Kenning, P. H., Davis, F. D., & Heekeren, H. R. (2014). Trusting humans and avatars: A brain imaging study based on evolution theory. *Journal of Management Information Systems, 30*(4), 83–114.

Tams, S., Hill, K., Ortiz de Guinea, A., Thatcher, J., & Grover, V. (2014). NeuroIS—Alternative or complement to existing methods? Illustrating the holistic effects of neuroscience and self-reported data in the context of technostress research. *Journal of the Association for Information Systems, 15*(10) (Article 1).

Vance, A., Anderson, B. B., Kirwan, C. B., & Eargle, D. (2014). Using measures of risk perception to predict information security behavior: Insights from electroencephalography (EEG). *Journal of the Association for Information Systems, 15*(10) (Article 2).

vom Brocke, J., & Liang, T.-P. (2014). Guidelines for neuroscience studies in information systems research. *Journal of Management Information Systems, 30*(4), 211–234 (Spring).

Appendix B: Major Statements in the NeuroIS Literature on the Importance of Cognitive Neuroscience Knowledge Acquisition

In the following, we present major statements in the NeuroIS literature on the importance of becoming familiar with the neuroscience literature in a given study context and the application of neuroscience knowledge in IS research without necessarily using neuroscience tools.

- Dimoka et al. (2007): "Our proposition for IS researchers is to first become familiar with the cognitive neuroscience literature and utilize its findings to enhance our understanding of IS phenomena ... We propose that many of the discoveries in cognitive neuroscience can have direct implications for IS theories. This can open up directions for research that could accelerate progress toward understanding the complex and elusive issues concerning the interplay of information, technology, and human behavior ..." (p. 1).

- Loos et al. (2010): "On many issues related to the perception (or creation) of artifacts, neuroscientific findings already exist that can be used ... This is especially worth mentioning as these results can inform the design of artifacts—without conducting any additional neuroscientific measurements" (p. 398). ... "Cognitive neuroscience is refining and advancing the theoretical foundations of traditional reference disciplines for IS research, including psychology, economics, and organizational behavior. It should be possible to take advantage of knowledge flowing from cognitive neuroscience for refining IS theories and hypotheses which can then be tested using traditional behavioral methods, without necessarily relying on brain measurements per se" (p. 400).

- Riedl et al. (2010): "Given the extensive neuroscience literature that has been developed during the last decades, there are many insights that can be drawn from the existing body of knowledge to inform IS research" (p. 249). ... "[B]enefiting from NeuroIS research does not necessarily imply conducting empirical neuroscience studies ... Rather, it is equally important to apply the knowledge that has already accumulated in the neuroscience literature to inform IS research questions. Against this background, it could be fruitful, for example, to (1) motivate future behavioral studies, (2) design behavioral experiments, (3) substantiate the

© Springer International Publishing AG 2017
R. Riedl et al., *Neuroscience in Information Systems Research*,
Lecture Notes in Information Systems and Organisation 21,
DOI 10.1007/978-3-319-48755-7_7

conclusions of behavioral investigations, and (4) even question existing assumptions ánd paradigms on the basis of neuroscience theories" (p. 249).

- Dimoka et al. (2011): "Challenge IS assumptions by identifying differences between existing IS relationships and the brain's underlying functionality, thus helping to build IS theories that correspond to the brain's functionality" (p. 688).
- Dimoka et al. (2012): "Because neurophysiological tools have the inherent attraction of being able to open the black box of the human brain and many studies have numerous interesting findings, there may be a rush among IS researchers to conduct NeuroIS studies without developing adequate knowledge of the neuroscience literature ..." (p. 695).
- vom Brocke et al. (2013): "This kind of theorizing illustrates the potential of neuroscience theories in building and evaluating IT artifacts in IS design science research without using neuroscience tools" (pp. 6–7). Note: Before this conclusion, vom Brocke et al. (2013) describe an example in the context of business process modeling notations which shows that neuroscience knowledge can be used to inform the building and evaluation of IT artifacts (see strategy 1 in their paper, p. 3). We stress that the formal logic implicitly underlying their example is consistent with our abstracted logic presented in Fig. 3.8 of the current book. This fact substantiates the efficacy of the approach presented in the current book.

Despite these statements on the importance of the application of neuroscience knowledge in IS research without necessarily using neuroscience tools, none of these papers delineates the exact nature of this neuroscience knowledge application process in the IS discipline. The order of the papers is based on date of publication.

References

Dimoka, A., Banker, R. D., Benbasat, I., Davis, F. D., Dennis, A. R., Gefen, D., et al. (2012). On the use of neurophysiological tools in IS research: Developing a research agenda for NeuroIS. *MIS Quarterly, 36*(3), 679–702.

Dimoka, A., Pavlou, P. A., & Davis, F. D. (2007). NEURO-IS: The potential of cognitive neuroscience for information systems research. *Proceedings of the 28th International Conference on Information Systems*, pp. 1–20.

Dimoka, A., Pavlou, P. A., & Davis, F. D. (2011). NeuroIS: The potential of cognitive neuroscience for information systems research. *Information Systems Research, 22*(4), 687–702.

Loos, P., Riedl, R., Müller-Putz, G., vom Brocke, J., Davis, F. D., Banker, R. D., et al. (2010). "NeuroIS: Neuroscientific approaches in the investigation and development of information systems. *Business and Information Systems Engineering, 2*(6), 395–401.

Riedl, R., Banker, R. D., Benbasat, I., Davis, F. D., Dennis, A. R., Dimoka, A., et al. (2010). On the foundations of NeuroIS: Reflections on the Gmunden Retreat 2009. *Communications of the Association for Information Systems*, 27, 243–264.

vom Brocke, J., Riedl, R., & Léger, P.-M. (2013). Application strategies for neuroscience in information systems design science research. *Journal of Computer Information Systems, 53*(3), 1–13.

Appendix C: Conceptual Description of Basic Brain Functioning from a Cognitive Neuroscience Perspective

Throughout the history of cognitive neuroscience, there has been an ongoing debate as to whether a mental process (e.g., trust) is localized in a discrete brain region or whether it is represented by a distributed network of brain regions (McIntosh 2000).

A developmental history of scientific knowledge of the brain might begin in the 19th century, with the pioneering research of phrenologists, who asserted that the human brain is organized around a finite number of specific mental processes, each of which was hypothesized to be located in a discrete brain region (Gazzaniga et al. 2009). Thus, phrenologists assumed a one-to-one mapping between a mental process and an anatomical region in the brain. This notion was referred to as *localizationist view*. As an example, Pierre P. Broca (1824–1880) and Carl Wernicke (1848–1905) localized the brain regions for speech production (the Broca's area) and speech comprehension (the Wernicke's area) through investigations of focal brain damage that resulted in speech deficits.[1]

However, as time progressed, researchers recognized that the brain cannot operate in a simple one-to-one fashion, particularly because the number of possible mental processes exceeds the number of brain regions (Price and Friston 2005).[2] Scholars began to acknowledge that even though discrete brain regions might underlie basic sensory and motor processes (McIntosh 2000), a network of regions and the interaction between them are crucial for the implementation of mental

---

[1]Recent research shows that the neural implementation of speech is based on a complex network of activity in distinct brain regions (e.g., Poeppel and Monahan 2008). However, although it is an established fact that speech is not based on activity in one or two brain regions (as most researchers assumed in the 19th century), current studies in brain science hold little doubt that the Broca and Wernicke areas are among the most important regions that are necessary for the neural implementation of speech production and comprehension (e.g., Kolb and Wishaw 2009, pp. 536–543).
[2]For example, the Brodmann areas atlas, one of the most influential atlases in neuroscience, divides the cerebral cortex into 52 areas on the basis of their cytoarchitecture (i.e., differences in cell types and their distribution).

© Springer International Publishing AG 2017
R. Riedl et al., *Neuroscience in Information Systems Research*,
Lecture Notes in Information Systems and Organisation 21,
DOI 10.1007/978-3-319-48755-7_8

processes such as trust (Gazzaniga et al. 2009). In this context, Cacioppo et al. (2008, p. 66) write: "[M]any of these [mental] processes involve a network of distributed, often recursively connected, interacting brain regions, with the different areas making specific, often task-modulated contributions. Moreover, a single neural region can often be involved in what have been treated as very different [mental] processes. One implication is that what have been considered basic psychological or behavioral processes are being conceptualized as manifestations of computations performed by networks of widely distributed sets of neural regions".

Currently, there is general agreement that mental processes arise from the activity of regions in a distributed network in the brain rather than from activity in one single brain region. As a consequence, a *network view*, rather than a localizationist view, dominates in cognitive neuroscience. This network view is of particular importance for NeuroIS research, because the constructs that are of interest to IS research are typically mental processes rather than basic sensory and motor processes. Examples are calculation, intentions, social cognition, trust, distrust, and moral judgment (see Dimoka et al. 2011, Table 3, p. 691, for a list of thirty-four mental processes).

Figure C.1 conceptually illustrates, in a simplified way, the neural implementation of a mental process that is based on the network view. Panel A shows a number of brain regions (open circles), each consisting of millions of neurons, with the anatomical connectivity between the regions—it is estimated that the human brain has a cranial capacity of approximately 1400 cm^3 and consists of 100 billion neurons. Moreover, the human brain is only semi-connected, due to anatomical constraints (McIntosh 2000); for example, each neuron is estimated to have connections to 10,000 other neurons. Therefore, not every single brain region has connections with all other regions.

Panel B illustrates activity in eight regions (black circles). This specific activity pattern exemplifies the neural implementation of a specific mental process. Panel C shows another activity pattern, thereby indicating the implementation of another mental process. However, the mental processes in Panel B and in Panel C share a common set of three brain regions—the three regions at the top of the network. Thus, a brain region may contribute to the implementation of more than one mental process. Considering this, and bearing in mind that a mental process is typically based on a pattern of activity in many brain regions rather than activity in one single region (Gazzaniga et al. 2009), the recent cognitive neuroscience literature posits a many-to-many mapping between mental processes and brain regions (Gonsalves and Cohen 2010; McIntosh 2000; Mesulam 1990; Price and Friston 2005). Finally, Panel D illustrates the implementation of another mental process based on activity in three brain regions. It is important to note that this mental process does not share any common regions with the other two mental processes illustrated in Panels B and C.

Altogether, Fig. C.1 shows the complexity of neural processes in the human brain, because (i) different mental processes may share common brain regions (see Panels B and C), but need not do so (see Panels B and D, as well as Panels C and D), and (ii) several brain regions are involved in the implementation of one specific mental process (see Panels B, C, or D).

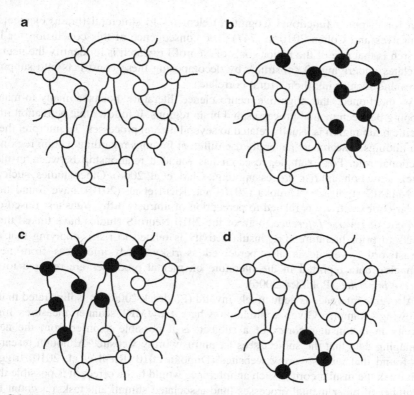

Fig. C.1 Neural implementation of mental processes based on the network view (conceptual illustration). The concept of this illustration is based on Fig. 1 in McIntosh (2000)

Against the background of the network view, attempts to localize a complex mental process in one specific brain region cannot be successful. Gonsalves and Cohen (2010, p. 747) write that "the problem is the dearth of theories that decompose complex psychological constructs into well-defined information processing stages or components ... i.e., a well developed *cognitive ontology* ... There are surely so many subprocesses associated with high-level psychological constructs that the results of fMRI studies will involve activity in many, widely distributed brain regions" (italics added).

Considering this argumentation, several cognitive neuroscience scholars have called for the development of cognitive ontologies (e.g., Henson 2005; Price and Friston 2005). A cognitive ontology is a taxonomy in the form of a hierarchy in which a mental process is decomposed into subprocesses, and each subprocess (rather than the complex mental process) is assigned to brain regions that are necessary for its neural implementation (Poldrack 2010; see also www. cognitiveatlas.org).

Consequently, the network view is *not* a move away from localization. Instead, it is a "modern view of localization" that informs scholars about the "right level of grain for mapping functional (cognitive) elements to structural (brain) elements" (Gonsalves and Cohen 2010, p. 747). The consequence of this conclusion for IS research is that one of the major goals of NeuroIS research is to identify the neural correlates underlying IS constructs by decomposing these constructs into subprocesses and identifying their neural correlates.

As mentioned, the cognitive neuroscience literature posits a many-to-many mapping between mental processes and brain regions. It follows that activation in a specific brain region is usually related to several mental processes. To interpret their own findings, researchers generally use different findings regarding a brain region's functional role. For example, research has found a relationship between insular cortex activation and risk processing (e.g., Clark et al. 2008). Other studies, such as the NeuroIS studies by Dimoka (2010) and Riedl et al. (2010), have found that insular cortex activity is related to perception of untrustworthy websites. Based on the logic of *reverse inference*, both of the 2010 NeuroIS studies have linked their results to prior literature (i.e., insula activity is related to risk), implying that an untrustworthy website is also perceived as risky. Such inferences from neuroimaging data reported in the literature on mental processes are referred to as *reverse inference* (Poldrack 2006).

A reverse inference is deductively invalid (Poldrack 2006), as is illustrated in the following example: Given a situation where the fMRI scanner measures high activity in the insular cortex of a subject, is it possible to infer from the neuroimaging data that the subject sees an untrustworthy website? Although research has found that untrustworthy websites (Dimoka 2010; Riedl et al. 2010) trigger activity in the insular cortex, such an inference would be incorrect. It is possible that a number of other mental processes (and associated stimuli and tasks) account for activity in this region (e.g., unfair offers in economic games, Sanfey et al. 2003, or faces expressing feelings of disgust, Phillips et al. 1997).

On the subject of reverse inference, Huettel et al. (2009) indicate that reasoning from brain activation in a specific region to a specific mental process means to take a dependent variable, the fMRI activation, to infer the state of an independent variable, the generating mental process. Moreover, they argue that the challenge of reverse inference can be distilled to a single concept, *selectivity*, which is a property of the dependent variable. fMRI activation (the dependent variable) is highly selective if it occurs only as a result of one mental process (or a very limited number of mental processes), whereas it has low selectivity if it could result from a large number of mental processes. Thus, high selectivity is positively correlated with justification of a reverse inference (Huettel et al. 2009; Poldrack 2006, 2010). Moreover, extending the mapping of specific and well-defined mental processes to increasingly smaller brain areas positively affects selectivity (Gonsalves and Cohen 2010). When very fine-grained mappings for specific mental processes of very small brain areas are available, the question arises about the configuration and temporal order of these mental processes, so that a higher-order construct can emerge. For example, how do the brain areas underlying reward, uncertainty, mentalizing, and learning interact in order to generate trust or distrust perceptions?

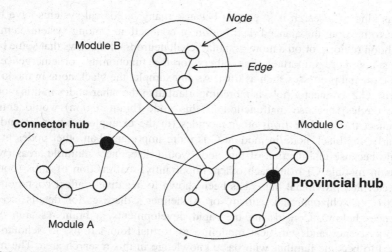

Fig. C.2 Network hubs in the brain (conceptual illustration). The concept of this illustration is based on Fig. 1 in van den Heuvel and Sporns (2013)

To the best of our knowledge, this kind of question has not yet been answered satisfactorily in cognitive neuroscience.

A recent extension of the network view deals with *network hubs* in the brain (for a review, see Sporns 2014). Such hubs are brain regions occupying a central position in the overall organization of a network that underlies the neural implementation of a specific mental process. Drawing upon network science and statistical modeling as theoretical lenses, cognitive neuroscientists have begun to investigate mental processes from this network perspective. Specifically, graph theory is used to formally describe brain networks, based on nodes (usually brain regions) and edges (usually synaptic connections). This principle of network hubs in the brain is conceptually illustrated in a simplified way in Fig. C.2.

Two different types of hubs can be distinguished: connector and provincial. According to van den Heuvel and Sporns (2013, p. 683), a connector hub is "a high-degree network node that displays a diverse connectivity profile across several different modules in a network" (see, for example, the black circle in module A of Fig. C.2) and a provincial hub is "a high-degree network node that mostly connects to nodes within its own module" (see, for example, the black circle in module C of Fig. C.2). Using different graph measures from network science (e.g., node degree, closeness centrality), together with knowledge of anatomical connections between brain regions, it has become possible to determine connector hubs and provincial hubs.[3]

---

[3]Node degree: the number of edges that are maintained by each node. Closeness centrality: the average distance (the length of the shortest paths) between a given node and the rest of the

This kind of research is important because many biological systems have been shown to have an unbalanced distribution of connections among system elements (here brain regions, or on a more granular level, neurons). It follows that some brain regions are more important than others for the functioning, or emergence, of specific mental processes such as trust. As an example, the black node in module A (see Fig. C.2, connector hub) is more important than the other nodes in this module (white circles) because malfunction of this node (brain region) would entirely disconnect this module from other modules (in the example modules B and C). Similarly, the black node in module C is more important than other nodes in this module because malfunction of this node would disconnect multiple areas (white circles in module C) from each other. Importantly, malfunction of hubs (both of connector and provincial hubs) has been shown to negatively affect cognition and health (e.g., schizophrenia, autism, or Alzheimer's disease; for details, see the references below). Considering the rapid developments in brain research using network science and statistical modeling as formal foundations, IS scholars are advised to become familiar with basic knowledge in this research area. The major reason for this recommendation is that most IS constructs constitute complex mental processes that are based on activity in multiple brain regions, and hence network hubs are likely to play a significant role in the development of a sound understanding of these mental processes.

IS scholars who are interested in this recent research domain are advised to read the following list of seminal papers: Bullmore and Sporns (2012), Bressler and Menon (2010), Friston (2011), Sporns (2013, 2014), Rubinov and Sporns (2010), and van den Heuvel and Sporns (2013). Also, a relatively recent book by Sporns (2011) summarizes important foundations and data on networks and network hubs in the human brain. Because this book is written for a general audience (rather than for brain experts), it may be a good starting point for IS researchers to become familiar with the network organization of the brain.

References

Bressler, S. L., & Menon, V. (2010). Large-scale brain networks in cognition: Emerging methods and principles. *Trends in Cognitive Sciences, 14*, 277–290.
Bullmore, E., & Sporns, O. (2012). The economy of brain network organization. *Nature Reviews Neuroscience, 13*(5), 336–349.
Cacioppo, J. T, Berntson, G. G., & Nusbaum, H. C. (2008, April). Neuroimaging as a new tool in the toolbox of psychological science. *Current Directions in Psychological Science, 17*(2), 62–67.

(Footnote 3 continued)

network. A number of additional graph measures can be found in Sporns (2011, Chap. 2) and Rubinov and Sporns (2010).

Clark, L., Bechara, A., Damasio, H., Aitken, M. R. F., Sahakian, B. J., & Robbins, T. W. (2008, May). Differential effects of insular and ventromedial prefrontal cortex lesions on risky decision-making. *Brain, 131*(5), 1311–1322.

Dimoka, A. (2010, June). What does the brain tell us about trust and distrust? Evidence from a functional neuroimaging study. *MIS Quarterly, 34*(2), 373–396.

Dimoka, A., Pavlou, P. A., & Davis, F. D. (2011, December). NeuroIS: The potential of cognitive neuroscience for information systems research. *Information Systems Research, 22*(4), 687–702.

Friston, K. (2011). Functional and effective connectivity: A review. *Brain Connectivity, 1*(1), 13–36.

Gazzaniga, M. S., Ivry, R., & Mangun, G. R. (2009). *Cognitive neuroscience: The biology of the mind* (3rd ed.). New York: W.W. Norton.

Gonsalves, B. D., & Cohen, N. J. (2010, November). Brain imaging, cognitive processes, and brain networks. *Perspectives on Psychological Science, 5*(6), 744–752.

Henson, R. (2005, February). What can functional neuroimaging tell the experimental psychologist? *Quarterly Journal of Experimental Psychology, 58*(2), 193–233.

Huettel, S. A., Song, A. W., & McCarthy, G. (2009). *Functional magnetic resonance imaging* (2nd ed.). Sunderland, MA: Sinauer Associates.

Kolb, B., & Wishaw, I. Q. (2009). *Fundamentals of human neuropsychology* (5th ed.). New York: Palgrave Macmillan.

McIntosh, A. R. (2000, November). Towards a network theory of cognition. *Neural Networks, 13* (8–9), 861–870.

Mesulam, M. M. (1990, November). Large-scale neurocognitive networks and distributed processing for attention, language, and memory. *Annals of Neurology, 28*(5), 597–613.

Phillips, M. L., Young, A. W., Senior, C., Brammer, M., Andrew, C., Calder, A. J., et al. (1997, October). A specific neural substrate for perceiving facial expressions of disgust. *Nature, 389* (6650), 495–498.

Poeppel, D., & Monahan, P. J. (2008, April). Speech perception: Cognitive foundations and cortical implementation. *Current Directions in Psychological Science, 17*(2), 80–85.

Poldrack, R. A. (2006, February). Can cognitive processes be inferred from neuroimaging data? *Trends in Cognitive Sciences, 10*(2), 59–63.

Poldrack, R. A. (2010, November). Mapping mental function to brain structure: How can cognitive neuroimaging succeed? *Perspectives on Psychological Science, 5*(6), 753–761.

Price, C. J., & Friston, K. J. (2005, June). Functional ontologies for cognition: The systematic definition of structure and function. *Cognitive Psychology, 22*(3–4), 262–275.

Riedl, R., Hubert, M., & Kenning, P. (2010, June). Are there neural gender differences in online trust? An fMRI study on the perceived trustworthiness of eBay offers. *MIS Quarterly, 34*(2), 397–428.

Rubinov, M., & Sporns, O. (2010). Complex network measures of brain connectivity: Uses and interpretations. *NeuroImage, 52*, 1059–1069.

Sanfey, A. G., Rilling, J. K., Aronson, J. A., Nystrom, L. E., & Cohen, J. D. (2003, June). The neural basis of economic decision-making in the ultimatum game. *Science, 300*(5626), 1755–1758.

Sporns, O. (2011). *Networks of the brain*. Cambridge, Massachusetts: MIT Press.

Sporns, O. (2013). Network attributes for segregation and integration in the human brain. *Current Opinion in Neurobiology, 23*, 162–171.

Sporns, O. (2014). Contributions and challenges for network models in cognitive neuroscience. *Nature Neuroscience, 17*(5), 652–660.

van den Heuvel, M. P., & Sporns, O. (2013). Network hubs in the human brain. *Trends in Cognitive Sciences, 17*(12), 683–696.

Appendix D: Description of Background Information on Online Trust

D.1 Why Did We Select Trust as an Example Topic?

First, two studies (Sidorova et al. 2008; Steininger et al. 2009) identified trust as one of the major topics in both North American and European IS research. Sidorova et al. (2008), investigating the intellectual core of the IS discipline, analyzed 1615 abstracts of articles published in three North American IS journals from 1985 to 2006, and found that trust is among the most important research topics in the time period 2002–2006, thereby demonstrating that the topic is up-to-date.[1] Steininger et al. (2009) investigated the titles, abstracts, and keywords of 5647 IS articles published in three North American and two European outlets from 1994 to 2007 and identified trust as the only trend in IS research among the fifty most important topics (a trend is defined as a topic that is addressed with permanently increasing intensity from 1994 to 2007).[2]

Second, there is reason to believe that trust has not only been a major theme in IS during the past two decades but will remain an important topic in the IS discipline in the future. This continued importance will be influenced, particularly, by the continued expansion of the role of the Internet, which is characterized by a high degree of anonymity, thereby increasing uncertainty perceptions of online users (Pavlou et al. 2007). The uncertainty, in turn, is strongly interrelated with trust (Pavlou 2003). Confirmation that trust will remain an important theme in IS research in the future is suggested by the special focus on "Trust in Online Environments" in the 2008 *Journal of Management Information Systems* (Vol. 24, No. 4), and is further substantiated by the 2010 *MIS Quarterly* special issue on "Novel Perspectives on Trust in Information Systems" (Vol. 34, No. 2).

---

[1]The three journals analyzed are *MIS Quarterly*, *Information Systems Research*, and *Journal of Management Information Systems*.
[2]The North American outlets analyzed are *MIS Quarterly*, *Information Systems Research*, and *Communications of the ACM*. The European outlets are *Information Systems Journal* and *Information & Management*.

© Springer International Publishing AG 2017
R. Riedl et al., *Neuroscience in Information Systems Research*,
Lecture Notes in Information Systems and Organisation 21,
DOI 10.1007/978-3-319-48755-7_9

Third, we have chosen trust because it is related to investigations on various levels of analysis in IS research, thereby demonstrating its ubiquity in and relevance for the IS community. Several research streams do exist in the trust literature, spanning the various levels of analysis in IS research: individual, group, organization, and society (Vessey et al. 2002). Notable streams of research on the individual level include investigations into trust in IT artifacts such as recommendation agents or avatars (e.g., Wang and Benbasat 2008) and trust in e-commerce (e.g., Pavlou and Gefen 2004). A major research stream that focuses on the group level is trust in virtual teams (e.g., Jarvenpaa et al. 2004). On the organizational level, trust in inter-organizational collaboration (e.g., Nicolaou and McKnight 2006) is of high relevance. Finally, on the societal level, trust in e-government (e.g., Teo et al. 2008) and trust in virtual worlds and communities (e.g., Johnson and Kaye 2009) are gaining considerable momentum.[3] Considering these multifarious research streams, trust is apparently a ubiquitous and relevant topic in IS research.

Fourth, and perhaps strongest of the reasons for choosing trust as an example topic, is that disciplines such as neuroeconomics, social neuroscience, and neuromarketing have several years ago identified trust as one of the most promising themes for investigation by means of neuroscience (e.g., Fehr and Camerer 2007; Kenning and Plassmann 2005). Hence, a rich body of knowledge on the brain mechanisms underlying human trust behavior does exist (e.g., Fehr 2009a, b; Riedl and Javor 2012). We call for an application of these existing literatures to advance IS theorizing.

In the present book, the focus of our discussion is *online trust* (for an overview of online trust, see Gefen et al. 2008). Although such an illustration is concentrated on the individual level of analysis, the implications of our discussion also pertain to the other levels of analysis (group, organization, society), particularly because, although brain activity always refers to the level of the individual, activity in a person's brain can nevertheless be influenced by, and can influence, behavior at the other levels (Riedl et al. 2010a).

D.2   Structure of an Online Trust Situation

The analysis of an online trust situation (e.g., buying a product on eBay) is typically focused on settings involving two specific parties: a trusting party (trustor) and a party to be trusted (trustee). Once a trustor (typically an online buyer) and a trustee (usually an online seller) begin to interact via the Internet (i.e., a seller puts a product on eBay and a buyer views the offer), the trustor can visually perceive stimuli on the screen, which provides rich information. This information may be

---

[3]The list of research streams is intended to outline important examples for the various levels of analysis in the trust literature; it is not intended to be exhaustive.

textual, graphical, or a blend of both. A sample eBay website demonstrates important information elements (see Fig. D.1): eBay logo, product name (New USB Flash Drive), picture of the product, selling mode (Buy It Now, indicating eBay's online shopping feature rather than a bidding auction), price (EUR 30.00), seller's name (usb-shop-123), seller's experience level (55, with a blue star indicating that 50 to 99 feedback ratings have been posted), feedback (100 % positive), duration and location of membership (since October 6, 2004, in Germany), and, finally, the product description.

A trustor processes information on a website to learn more about the product for sale. However, the information is also used to infer the trustee's characteristics, thereby making possible predictions regarding his/her trustworthiness (Riedl et al. 2010b; Riegelsberger et al. 2005). From the viewpoint of the trustee, information provision on a website serves the purpose of deliberately signaling trustworthiness. Important characteristics of a trustee, which together constitute a trustee's trustworthiness, are ability, benevolence, and integrity (Gefen 2002; Mayer et al. 1995). If a trustor believes in the trustworthiness of a trustee, he/she believes that the trustee (i) has skills and competencies that are important for the relationship

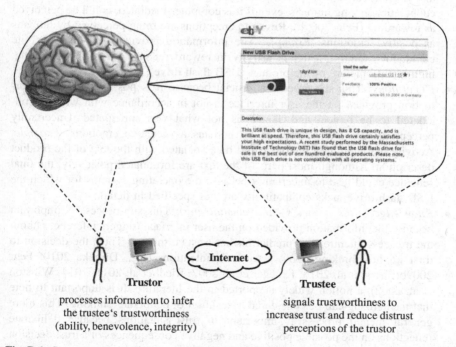

Fig. D.1 Structure of an online trust situation (example: eBay). The eBay website in the illustration is taken from Riedl et al. (2010b, p. 408)

(ability), (ii) wants to do good to the trustor, aside from an egocentric profit motive (benevolence), and (iii) adheres to a set of principles that the trustor finds acceptable (integrity). The eBay website illustrated in Fig. D.1, for example, shows that the trustee's feedback profile contains 100 % positive feedback. This information could be used by the trustor to evaluate the trustee's degree of benevolence. However, all other available information elements on the website may also be used to determine the trustee's ability, benevolence, and integrity.

From a cognitive neuroscience viewpoint, it is interesting to study the neural effects caused by the perception and processing of the information provided on a website. Engaging in a trust decision has been demonstrated to be a trigger for activity in several regions in the trustor's brain (Baumgartner et al. 2008; Delgado et al. 2005; Dimoka 2010; King-Casas et al. 2005; Krueger et al. 2007; Riedl et al. 2010b; Winston et al. 2002). The specific brain areas are related to the following four major constructs (for a review, see Riedl and Javor 2012; all examples refer to the eBay website in Fig. D.1):

- *Reward*: This sub-process is important because the outcome of an Internet shopping process, the product purchased, is expected to provide value—a value that is often anticipated in the brain before using the product. Moreover, social cooperation among humans, even if it is computer-mediated, could be perceived as rewarding (Fehr 2009b). Reward perceptions are often perceived by humans as positive emotions. *Example*: The information provided on the eBay site could, for example, result in activity in reward regions due to an appealing, high-quality picture of the product (USB flash drive).

- *Uncertainty*: This sub-process is crucial because it is possible that the information provided on the user interface is not in accordance with what is considered to be typical and, hence, is not what was anticipated. Uncertainty perceptions are often perceived by humans as negative emotions. *Example*: Activity in uncertainty regions could be associated with the text of the product description. Although most parts of the text are formulated positively, the final sentence could lead to uncertainty because the operating systems for which the USB flash drive lacks compatibility are not specified in detail.

- *Knowledge of the trustee's mind (mentalizing)*: This sub-process is important because the information provided on the user interface triggers inferences about the trustee's intentions to predict his/her trustworthiness. Thus, the decision to trust implies thinking about a trustee's intentions (e.g., Dimoka 2010; Fehr 2009b; Javor et al. 2016; Krueger et al. 2007; Riedl et al. 2011, 2014; Winston et al. 2002), a notion widely supported in the literature. It is important to note that mentalizing can be considered as a subset of deliberate thinking. This more general mental process is important in trust situations because deliberate reflections on the possible positive and negative consequences of a trust decision play a crucial role in complex social interactions. *Example*: Based on a deliberate integration of all the information provided on the eBay website, a trustor might reflect on the possible consequences of both trust and distrust behavior in order to arrive at a final decision.

- *Learning*: Prominent IS researchers (e.g., Gefen et al. 2008) have called for more investigations into the longitudinal nature of trust. Because trust develops gradually as people interact with each other, the longitudinal effects of trust on behavioral outcome variables (e.g., transaction intention) are affected by interaction experience. This experience may relate to transactions with a particular agent, or to online transactions in general. Accentuating the longitudinal nature of trust implies that a trustor can learn about the behavior of trustees. *Example*: Based on past transactions with a specific seller, an eBay buyer could know that this seller is trustworthy.

D.3 A Conceptual Framework for Trust in Online Environments

In this section, we present a conceptual framework for trust in online environments. Figure D.2 puts the four constructs associated with trust into a logical and temporal order.

A look at our conceptual framework for trust in online environments (Fig. D.2) reveals that a trust decision involves various brain regions. Automatic, emotional, and unconscious information processing plays a crucial role for trust decisions. Prominent scholars such as Damasio (2005) and Zak (2003) support the importance of automatic, emotional, and unconscious information processing in trust situations; Zak (2003, p. 21), for example, provides a clear-cut statement: "[T]rust is not a calculative activity (i.e. determining costs and benefits), but a visceral sense that one has that a person can be trusted or not." However, other researchers have argued that trust also implies calculative, conscious, and rational thought processes (see, for example, mentalizing in our conceptual framework).[4]

In the following, we describe our conceptual framework (Fig. D.2). An IT artifact—for example, a user interface such as an online shopping site—consists of different information elements such as text, human images, and avatars. Once a user

---

[4]Coleman (1990), for example, conceptualizes a trust decision more as a calculative, conscious, and rational process. Based on his model, trust is defined as: $p \times G > (1 - p) \times L$. Notation: p, the likelihood of making a profit (gain), and this equals the probability that the trustee behaves in a trustworthy manner; L, the potential loss that occurs if trust is betrayed; and G, the potential gain if the trustee in fact turns out to be trustworthy. Fehr (2009b, p. 228), for example, writes that "[s]ince trust decisions are also likely to involve perspective-taking, they should also activate areas implicated in theory-of-mind [mentalizing] tasks".

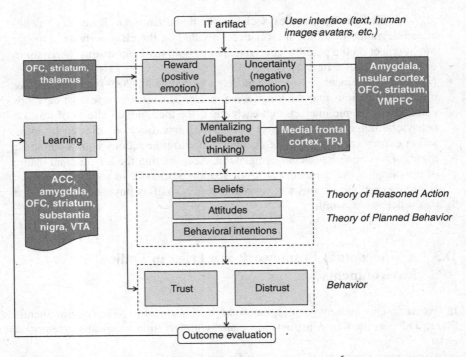

Fig. D.2 Conceptual framework for trust in online environments.[5] *Notes* The illustrated mechanism [Beliefs → Attitudes → Behavioral Intentions] is based on the Theory of Reasoned Action and Theory of Planned Behavior (Ajzen 1991; Fishbein and Ajzen 1975). The striatum consists of caudate nucleus, nucleus accumbens, and putamen. *ACC* Anterior Cingulate Cortex, *OFC* Orbitofrontal Cortex, *TPJ* Temporo-Parietal Junction, *VMPFC* Ventromedial Prefrontal Cortex, *VTA* Ventral Tegmental Area. We assigned *major* brain areas to the four sub-processes of trust. It follows that brain imaging studies investigating the neural correlates of reward, uncertainty, mentalizing, and learning have revealed further neural correlates; however, they are less important when compared to the described brain regions as discussed in the papers below (see section "The Four Major Sub-processes of Trust")

is confronted with such an interface, she/he begins to (visually) perceive these stimuli. Perceptions are then automatically compared to existing knowledge, and are typically followed by rapid and unconscious reward or uncertainty perceptions. Reward processes are perceived as positive emotions, while uncertainty processes

---

[5]As indicated by one reviewer, the predictive power of the conceptual framework in Fig. D.2 could be tested in future studies. Considering the argument of the present paper, the testing does not necessarily need to be carried out on the neural level alone. Rather, it is possible to test at least some of the relationships on a behavioral level also. As an example, the strong positive or negative emotions that may have direct influence on trust or distrust behavior (without the mediation via mentalizing and subsequent beliefs, attitudes, and behavioral intentions) could be tested in a behavioral study. The hypothesis would be that stimuli (e.g., websites) that induce either very positive or very negative emotions will more directly drive behavior (e.g., mouse click behavior on websites) than less emotional stimuli will. Directness could be measured based on reaction time.

are typically perceived as negative emotions. Importantly, very strong positive and negative emotions may directly result in automatic behavior, here rapid trust or distrust decisions.[6]

Next, neural processes may enhance attention (Shimojo et al. 2003) which, in turn, may lead to adjustments of the original reward or uncertainty perceptions. Increased attention levels may also affect more deliberate and conscious information processing. In social interaction situations (particularly, in trust situations), mentalizing is considered to be a major type of deliberate thinking (Dimoka 2010; Fehr 2009b). Altogether, the described neural processes lead to trust or distrust behavior. Finally, as a consequence of the success of the transaction (i.e., the transaction flows smoothly or does not), a trustor will learn on the basis of these experiences (outcome evaluation). This learning process may change future information processing, as well as future judgment and decision making, in trust situations (see the loop in the conceptual framework).

Based on our review of the cognitive neuroscience literature, we assign brain areas to the four constructs associated with trust in our conceptual framework. Our framework illustrates the many-to-many mapping between mental processes and brain regions (Price and Friston 2005).

D.4   Four Major Sub-processes of Trust

We focus the review in this section on four major constructs (reward, uncertainty, mentalizing, and learning) that are closely associated with human trust (for a detailed review, see Riedl and Javor 2012). Note that the main goal of this review is to highlight the diversity in stimuli and tasks that have been used to investigate the neural correlates of each construct in non-IS contexts; it is not our goal to provide a comprehensive review. In the final paragraph of the discussion of each construct,

---

[6]In an experiment, Willis and Todorov (2006) manipulated exposure time to unfamiliar and emotionally neutral faces in order to investigate judgments of trustworthiness. Judgments made after a 100 ms exposure correlated highly with judgments made in the absence of time constraints, and increased exposure times (from 100 to 500 ms and from 500 to 1000 ms) did not significantly increase the correlations. Thus, this experiment suggests that an exposure time of 100 ms is sufficient for humans to form a trustworthiness impression. Because exposure times of 100 ms or shorter are typically not sufficient for saccadic eye movements, trustworthiness impressions are hypothesized to be "single glance" impressions rather than impressions based on visual exploration of a face (Todorov 2008). This result is in line with our theorizing (as illustrated in Fig. D.2), which indicates that at stimulus onset, a person's brain may automatically and unconsciously form judgments (either positive or negative), which may then directly influence a trust or distrust decision.

we indicate seminal references—both review and meta-analysis papers—that we consider to be a valuable source of further knowledge consolidation.

Before we begin our review, it is important to stress that the four discussed research streams have different lineages. Whereas reward and learning have primarily been the domain of animal researchers, neuropsychologists, and economists, mentalizing has predominant roots in social and developmental neuroscience. Moreover, uncertainty (including risk and ambiguity) has been studied extensively in neuropsychology, cognitive neuroscience, and economics. It follows that readers should not perceive an incorrect impression that the research on the four constructs constitutes a coherent field. Rather, each literature has its own jargon and the respective researchers often do not communicate with each other. One of the exciting aspects of the potential of NeuroIS is the degree to which these and other fields need to come together to inform IS.[7]

D.5 Reward

In cognitive neuroscience, the concept of reward has two meanings (Schultz 2006). On the one hand, it relates to subjective feelings of pleasure and liking, the hedonic function of a reward. On the other hand, it is considered to be an object or event that is received for having done something well. In this section, we primarily refer to the former meaning. The latter meaning is addressed in more detail in the section on learning, as this specific notion of reward is closely associated with the paradigms of Pavlovian (Classical) Conditioning and Operant (Instrumental) Conditioning (Gazzaniga et al. 2009).[8] Despite this basic distinction, however, both meanings cannot be entirely separated.

The crucial role of reward for the survival and wellbeing of humans ranges from the control of vegetative functions to the organization of complex goal-directed behaviors (Schultz 2000). In order to survive, humans must drink, eat, and reproduce. Hence, drinks, food, and sex are major *primary rewards* (also referred to as basic rewards). However, as human societies have become more complex as they developed, *secondary rewards* (also referred to as cognitive rewards) have become increasingly more important—as in the examples of money, novelty, challenge, beauty, and power (Schultz 2000).

Although the exact brain mechanisms underlying reward perception and processing are not yet fully understood, literature reviews (O'Doherty 2004; Schultz

---

[7]We thank one of the reviewers for this comment. Moreover, we recommend reading a paper by Levallois et al. (2012), who investigated how interactions between neuroscience and social science have contributed to the nature of neuroeconomics research.

[8]Pavlovian (Classical) Conditioning: Learning a predictive relationship between a stimulus and a reinforcer (does not require an action by the subject). Operant (Instrumental) Conditioning: Learning a relationship between a stimulus, an action, and a reinforcer, conditional upon an action by the subject (Schultz 2000).

2006) suggest that the neural correlates of secondary rewards closely resemble those of primary rewards. In line with this finding, it has been argued that basic and cognitive rewards serve similar functions—they (i) induce subjective feelings of pleasure and contribute to positive emotions, (ii) act as positive reinforcers by increasing the frequency and intensity of behavior that results from the acquisition of goal objects and the establishment of desired states, (iii) maintain learned behavior by preventing extinction, and (iv) act as goals and can thereby elicit approach and consummatory behaviors (Schultz 2000).

To date, research based on various stimuli and experimental tasks (e.g., sight, hearing, taste, and smell) has identified the brain structures associated with the perception and processing of rewards (O'Doherty 2004). One fMRI study (O'Doherty et al. 2003), for example, operationalized reward based on the attractiveness of human faces. The task was to view 48 faces in random order (four repetitions, for a total of 192 face presentations). Following the presentation of each face, participants were instructed to press one of two buttons to indicate whether the face was that of a male or a female. An assumption in this experiment is that a heightened perception of attractiveness comes with higher reward. The study found that attractive faces, in contrast to unattractive ones, activated the orbitofrontal cortex (OFC), leading to the conclusion that OFC activity is associated with reward perceptions and/or reward processing.

Other investigations based on different stimuli and tasks found two further brain structures that play a significant role in reward perception and processing. First, the striatum, with its sub-components caudate nucleus and putamen, has been shown to be associated with reward (Schultz 2006; Schultz et al. 1997; Tricomi et al. 2004). This result is supported by the fact that dopamine, a neurotransmitter, is released by a specific part in the striatum, the nucleus accumbens, as a result of naturally rewarding experiences (e.g., Schultz et al. 1997).[9] Increased dopamine levels, in turn, positively affect activity in the striatum, which then leads to positive feelings and thereby reinforces and motivates approach and consummatory behaviors. Second, activity in the thalamus is also associated with reward. One study (Komura et al. 2001), for example, found that single thalamic neurons can code for the acquired significance of sensory stimuli, and can predict future reward value. In line with this result, another paper (Glimcher and Lau 2005) argues that the thalamus is part of the striatal loop and reflects information about reward magnitudes.

Altogether, cognitive neuroscience research has identified the OFC, striatum, and thalamus as major brain areas for reward perception and processing. A detailed description of the exact neurobiological mechanisms underlying reward perception and processing is available in the following review papers: Delgado and Rigney (2009), Doya and Kimura (2009), Knutson et al. (2009), O'Doherty (2004, 2009), Schultz (2000, 2002, 2006, 2009), and Smith and Delgado (2015).

---

[9]A neurotransmitter is a chemical that transmits signals from a nerve cell (i.e., a neuron) to a target cell across a synapse. This process influences the electrical activity of the target cell, thereby contributing to the transmission and processing of information within the brain (Kolb and Wishaw 2009, p. 84).

D.6   Uncertainty

In the literature on judgment and decision making, both in the behavioral and the brain sciences, uncertainty is often either conceptualized as risk or as ambiguity (Ellsberg 1961; Knight 1921; Platt and Huettel 2008). Risk is defined as uncertainty regarding which of several possible outcomes will occur, whereby the probability of each possible outcome is known. For example, a person who has to choose between two options—$10 with a probability of obtaining it of 0.50 or $6 with a probability of 0.80—is confronted with a risky decision. In contrast, in the case of ambiguity the possible outcomes of a decision, and/or their corresponding probabilities, are unknown. Thus, the amount of information a decision-maker has in risky situations is higher than in ambiguous situations. In real-life situations, many decisions are ambiguous rather than risky.

One paradigm that has been used frequently to study the neural correlates of uncertainty are gambles. In one study (Hsu et al. 2005), participants had to make choices between certain amounts of money and bets on card decks—and by experimental design the card decks reflected varying levels of risk and ambiguity. The study revealed that the brain region that was more active during the risk condition was the striatum. In contrast, in the ambiguous condition the OFC and the amygdala were more active. Moreover, a positive correlation between striatal activity and expected reward was found in the risk condition (expected reward is defined as outcome × probability). This result supports the striatum's crucial role in the processing of cognitive rewards.

With respect to the temporal order of brain activity, research indicates that the ambiguity-related regions (OFC and amygdala) react rapidly at the onset of stimuli (Hsu et al. 2005), whereas activation in areas related to the processing of the risk associated with a reward (e.g., striatum) builds more slowly (Preuschoff et al. 2006). Thus, cognitive neuroscience evidence suggests the following theoretical mechanism: When confronted with an ambiguous situation, the human brain is alerted, due to the fact that information is missing. The bidirectionally connected brain areas of the OFC and amygdala are crucial in performing this alert function, because they work together in evaluating the first impressions of stimuli (Hsu et al. 2005). If the first impression of an option is associated with strong negative perceptions (as is the case when the degree of ambiguity is very high), OFC and amygdala activation directly drive behavior (e.g., distrust, avoidance, and withdrawal behavior). However, OFC and amygdala activation can also induce information-seeking behaviors in order to find additional information from the environment, thereby transforming an ambiguous situation into a risky one (Hsu et al. 2005). This is consistent with behavioral research proposing that humans avoid ambiguity through increased information search (e.g., Frisch and Baron 1988), a behavior that is typically mediated by increased attention (Fiorillo et al. 2003).

In addition to the striatum, the OFC, and the amygdala, cognitive neuroscience studies based on gamble paradigms have identified additional brain regions that are associated with uncertainty perception and processing, namely the insular cortex (Kuhnen and Knutson 2005; Preuschoff et al. 2008), nucleus accumbens (NAcc, Kuhnen and Knutson 2005), anterior cingulate cortex (ACC, Critchley et al. 2001), ventromedial prefrontal cortex (VMPFC, Clark et al. 2008; Plassmann et al. 2008), and thalamus (Mohr et al. 2010).

The Kuhnen and Knutson (2005) study, for example, investigated whether neural activity would predict optimal and suboptimal choices in a financial decision-making task (gamble paradigm). Two types of deviations from the optimal investment strategy of a rational risk-neutral agent were defined by the experimenters: risk-seeking mistakes and risk-aversion mistakes. The results of the study show that NAcc activity preceded risky choices and risk-seeking mistakes, while insular cortex activity preceded riskless choices and risk-aversion mistakes. These findings provide neurobiological evidence that distinct neural structures associated with anticipatory affect promote different types of financial choices. Specifically, NAcc activity made the participants more risk-seeking, while insular cortex activity made the subjects more risk-averse. Altogether, the results substantiate the notion that excessive activity in reward-related structures (such as the NAcc) and in structures associated with uncertainty (such as the insular cortex) may result in suboptimal choices.

The practical implications of these research results are far-reaching. Kuhnen and Knutson (2005, p. 768) present two meaningful examples in their study: "This may explain why casinos surround their guests with reward cues (e.g., inexpensive food, free liquor, surprise gifts, potential jackpot prizes)—anticipation of rewards activates the NAcc, which may lead to an increase in the likelihood of individuals switching from risk-averse to risk-seeking behavior. A similar story in reverse may apply to the marketing strategies employed by insurance companies."

In addition to gamble paradigms, other stimuli and tasks have been used in studies of the neural correlates of uncertainty. Research has found that insular cortex activation is associated with the anticipation of emotionally aversive visual stimuli such as spiders and snakes (Simmons et al. 2004), as well as the anticipation of physical pain (Ploghaus et al. 1999). Moreover, insular cortex activation correlates with the perception of faces that express a feeling of disgust (Phillips et al. 1997, 1998). With respect to the perception of faces that express negative emotions (in particular, fear), it was found that the amygdala also plays a key role. For example, case studies of patients who had complete bilateral amygdala damage found that these patients judged other people to look more trustworthy and more approachable than did normal viewers, or patients with brain damage in other areas (Adolphs et al. 1998, 2005). This result suggests that the amygdala plays a crucial role for the processing of uncertain, arousing, and untrustworthy stimuli, especially human faces, which could turn out to be dangerous (e.g., Engell et al. 2007; Todorov 2008; Todorov et al. 2008). In line with this notion, Bechara and Damasio (2005, p. 353) state that the amygdala has evolved for a survival purpose—to be responsive to dangerous animals or to persons with untrustworthy faces, for example.

Altogether, cognitive neuroscience research across various experimental paradigms has identified the insular cortex (Singer et al. 2009), as well as the OFC and the amygdala (Hsu et al. 2005), as major brain areas for the perception and processing of uncertainty. Moreover, the striatum (Kuhnen and Knutson 2005) and the VMPFC (Clark et al. 2008) are also associated with uncertainty—as structures that are activated in risky situations (typically investigated based on gamble paradigms), in which expected rewards are computed based on outcome and probability information. More detailed descriptions of the neurobiological mechanisms underlying uncertainty perception and processing are available in two meta-analyses (Krain et al. 2006; Mohr et al. 2010), as well as in two review articles (Platt and Huettel 2008; Schultz et al. 2008).

D.7 Mentalizing

The ability to infer the internal states of other actors in order to predict their behavior is known as theory-of-mind (TOM), and the underlying inference process is commonly referred to as mentalizing (e.g., Frith and Frith 2003; Premack and Woodruff 1978; Singer 2009). The investigation of the neural mechanisms underlying mentalizing has gained considerable momentum during the past years, because the ability to make inferences about the intentions, feelings, and thoughts of other people is a unique characteristic that distinguishes humans from animals, and is, therefore, the basis for culture and civilization (Adolphs 2009).

One of the oldest experimental designs to investigate mentalizing is "the chocolate story," which constitutes a false-belief paradigm (Amodio and Frith 2006; Perner and Lang 1999; Singer 2009; Wimmer and Perner 1983). This story was developed to find out the age at which children begin to develop mentalizing abilities. In this experiment, children are told the following story. Max goes to the kitchen and puts his chocolate in the green cupboard. Then Max goes to the playground. Meanwhile, Max's mother comes into the kitchen and transfers the chocolate from the green to the blue cupboard. Afterwards, the mother leaves the kitchen, and Max returns from the playground and goes into the kitchen to get his chocolate. After this story has been told to a child, the following question is asked: Where will Max look for his chocolate? Obviously, a child who states that Max will look in the green cupboard knows that he falsely believe the chocolate to be there (because Max was not in the kitchen when his mother transferred the chocolate to the blue cupboard, so that he cannot know that the chocolate is no longer in the green cupboard).[10]

In order to investigate the neural correlates of mentalizing in adults, several experimental paradigms have been developed (Amodio and Frith 2006). One such

---

[10]At age six, almost all healthy children give the correct answer (Singer 2009).

paradigm, for example, is based on story comprehension. Here, participants perform story comprehension tasks necessitating the attribution of mental states, while their brain activity is being simultaneously measured. The following story serves as an example for a task that triggers mentalizing (Fletcher et al. 1995, p. 124): "A burglar who has just robbed a shop is making his getaway. As he is running home, a policeman on his beat sees him drop his glove. He doesn't know the man is a burglar, he just wants to tell him he dropped his glove. But when the policeman shouts out to the burglar, 'Hey, you! Stop!', the burglar turns round, sees the policeman and gives himself up. He puts his hands up and admits that he did the break-in at the local shop. Why did the burglar do this?" Obviously, reading this story and finding an answer to the question triggers mentalizing. The brain imaging results of the Fletcher et al. (1995) study show that, in contrast to control tasks, the story comprehension task led to significantly higher activation in the medial frontal gyrus.

In another brain imaging study (Gallagher et al. 2002), participants played a computerized version of the game "rock-paper-scissors." In the first experimental condition, denoted as *mentalizing*, participants believed they were playing against another human (although, in fact, they were playing against a random selection strategy). In the second condition, denoted as *rule solving*, participants were informed that they were playing against a computer with a predefined algorithm, and that the responses would be based on simple rules related to the participant's previous response (e.g., the computer would select the response that would have beaten the participant's last response). In the third condition, denoted as *random selection*, participants were informed that they were playing against a computer with a random selection strategy, and they were asked to respond randomly, as well. The contrast between mentalizing and rule solving revealed higher activation in the medial prefrontal cortex. Moreover, the contrast between mentalizing and random selection identified higher activation not only in the anterior paracingulate cortex, but also in the right inferior frontal cortex and the cerebellum. No activation was seen in the paracingulate cortex, however, when rule solving and random selection were compared. Because it was only in the mentalizing condition that participants believed they were playing against another human (who would have beliefs, desires, and intentions to interpret and predict the behavior of others), the bilateral anterior paracingulate cortex is considered to be a crucial mentalizing brain region (Singer 2009).

Finally, a study (Coricelli and Nagel 2009) investigated how a player's mental processing incorporates the thinking process of others in strategic reasoning. In the competitive interactive setting of an economic game, playing against human opponents (relative to playing against a computer) activated regions associated with mentalizing, specifically the medial prefrontal cortex and the rostral part of the ACC, which both jointly correspond approximately to the paracingulate cortex. This result supports the findings of previous experiments that are based on other paradigms (e.g., story comprehension).

Altogether, cognitive neuroscience research across various experimental paradigms has identified the medial frontal cortex as a major mentalizing brain region (Amodio and Frith 2006). Moreover, several studies have identified the temporo-parietal junction (TPJ) as significant in mentalizing situations (for details, see Saxe and Kanwisher 2003). Importantly, mentalizing is a subset of deliberate thinking, which is neurologically implemented primarily in the frontal cortex (Gazzaniga et al. 2009). A detailed description of the neurobiological mechanisms underlying mentalizing is available in various review articles (Abu-Akel 2003; Amodio and Frith 2006; Frith and Frith 2003, 2006, 2010; Gallagher and Frith 2003; Gallese 2007), as well as in two meta-analyses (Denny et al. 2012; Van Overwalle 2009).

D.8 Learning

Learning is acquiring new information or modifying existing information, and this may lead to changes in important variables such as knowledge, skills, and behaviors (Gazzaniga et al. 2009). Neuroimaging has made possible a direct assessment of changes in the brain when a specific task is learned. One pioneering study (Haier et al. 1992), for example, measured brain activity in subjects playing the computer game Tetris at two stages—before and after practice with the game. After several weeks of daily practice on Tetris, subjects' brain activity decreased significantly, despite a more than seven-fold increase in Tetris performance (which was measured based on the number of lines of the game completed). This finding supports the notion that learning leads to a more efficient use of the brain, reflected by a decrease in brain metabolism.

This example study by Haier et al. (1992) refers to *skill learning*. Here, cognitive and motor skills (e.g., thinking about falling blocks and using the keyboard to rotate and move them) are learned through practice, and can thereby be executed in an increasingly automatic manner. However, the following discussion focuses on *reward learning*.

Considering the unstable nature of interactions among humans, both in face-to-face settings and in online environments, it is not surprising that specific neural structures have evolved that not only make possible the identification of rewarding stimuli, but that also predict their occurrence based on past experiences. Importantly, the identification of accurate predictors for rewards is a major goal of learning (Fiorillo et al. 2003). Reward uncertainty, however, implies that a person lacks such accurate predictors. Thus, in the presence of uncertainty, motivation to find accurate predictors emerges, thereby prompting attention. The maximum amount of information is available for a reward with a probability of $p = 0.5$ (i.e., in the situation with the maximum degree of uncertainty), whereas rewards with $p = 1$ (no uncertainty) or $p = 0$ (no reward) do not contain new information.

Processing reward information in the presence of uncertainty (i.e., $0 < p < 1$) is typically associated with activation in dopamine neurons, which results in prediction error (δt) signals (see Formula 1). Such signals are a crucial part of reinforcement learning models (Rescorla and Wagner 1972), which since the 1990s have been intensively investigated by means of neuroscience (Niv and Montague 2009).

Reinforcement learning is learning by interacting with one's environment. In this paradigm, humans learn from the consequences of their actions (and not from being taught by other humans or by observing their behavior). Thus, future actions are chosen on the basis of past experiences.[11] The reinforcement signal that humans receive is a reward, which encodes the success of an action's outcome. People's objective is to learn to choose actions that maximize the accumulated reward over time.[12]

In a simple form, reinforcement learning models have the following structure (Behrens et al. 2009):

$$Vt + 1 = Vt + \delta t \times \alpha t \tag{1}$$

This model indicates that expectations of future reward ($Vt + 1$) are a function of current expectations (Vt) and their deviation from the experienced outcome, the prediction error (δt). Moreover, the model considers a learning rate (αt). This ensures that future expectations can be updated by the product of the prediction error and the learning rate.

To date, a number of neuroscience studies investigated the brain regions underlying the parameters in reinforcement learning models. In essence, the striatum, OFC, and amygdala encode expectations of future reward, as well as current reward expectations (e.g., Gottfried et al. 2003; Schultz 2002). Activity in the ventral tegmental area (VTA) and the substantia nigra is associated with the reward prediction error (e.g., Behrens et al. 2009; Niv and Montague 2009; O'Doherty et al. 2003; Schultz et al. 1997).

A prediction error can be positive or negative (e.g., O'Doherty et al. 2004). Hence, the unexpected presentation of a reward leads to an increase in brain activity in specific areas, while unexpected omission of a reward results in a decrease of activity in specific areas. One fMRI study (Yacubian et al. 2006) found that positive prediction errors are associated with activation in the striatum, while negative

---

[11]In the case of completely new environments, an explorative strategy such as trial and error learning is pursued.

[12]See Scholarpedia (reinforcement learning).

prediction errors are associated with activity in the amygdala. Moreover, one investigation (Wheeler and Fellows 2008) found that the ventromedial frontal lobe is critical for learning from negative feedback. Thus, the results of these studies suggest that the striatum, amygdala, and ventromedial frontal lobe compute a prediction error for gains and losses, respectively. As a consequence, the normal functioning of these brain structures is crucial for the accurate functioning of reward learning mechanisms. The prevalence of either structure might result in too positive (striatum) or too negative expectations (amygdala and ventromedial frontal lobe), which might lead to sub-optimal decisions.

The remaining component in Formula 1 is the learning rate. This component relates to the issue of how rapid the learning from new experiences should be, and how much existing knowledge should be preserved. The appropriate choice of the learning rate depends on both the structure of the environment and the experience of the agent (Doya 2008). In a constant environment, the best way to learn is to begin with rapid memory updating, followed by a reduction of the learning rate as an inverse of the number of experiences. In a dynamic environment, in contrast, the learning rate should be linked to the time for which the validity of past experiences remains unchanged. Cognitive neuroscience research has found that activity in the ACC is related to the learning rate in reinforcement learning paradigms (Behrens et al. 2007; Rushworth et al. 2007).

Altogether, cognitive neuroscience research across various experimental paradigms has identified, in particular, the striatum, OFC, VTA, and substantia nigra as reward learning brain regions. Moreover, limbic structures, the amygdala (processing of negative prediction errors), and the ACC (associated with the learning rate) also play a role in the learning of rewards. More detailed descriptions of the neurobiological mechanisms underlying reward learning are available in several review articles (Daw and Doya 2006; Dayan et al. 2009; Niv and Montague 2009; O'Doherty 2004; Schultz 2006), as well as in two meta-analyses (Chase et al. 2015; Garrison et al. 2013).

D.9 Summary

Table D.1 summarizes selected findings from the cognitive neuroscience literature. The review reveals, among other things, that each construct is associated with activity in multiple brain areas, and different constructs are based on activity in a partly common set of brain areas (e.g., uncertainty and learning are associated with activity in the ACC, OFC, and striatum). Moreover, the review shows that some

Table D.1 Summary of major neural correlates of the four constructs

Brain region	Reward	Uncertainty	Mentalizing	Learning
ACC			•	•
Amygdala		•		
MFC			•	
Insular cortex		•		
OFC	•	•		•
Striatum	•	•		•
Substantia nigra				•
TPJ			•	
Thalamus	•			
VMPFC		•		
VTA				•

Locations in the brain

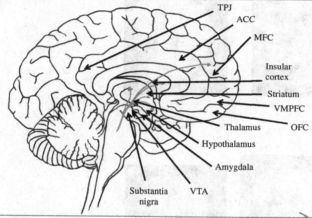

Notes ACC Anterior Cingulate Cortex, *MFC* Medial Frontal Cortex, *OFC* Orbitofrontal Cortex, *TPJ* Temporo-Parietal Junction, *VMPFC* Ventromedial Prefrontal Cortex, *VTA* Ventral Tegmental Area

brain regions are associated with several constructs (e.g., the OFC and striatum are associated with three constructs in each case), while other regions (e.g., VMPFC and VTA) are related to fewer constructs.

In addition to our narrative description of major studies in the four construct domains, we used *Neurosynth* (an online platform for large-scale, automated synthesis of fMRI data, http://neurosynth.org/) to quantitatively analyze studies that investigate the neural correlates of reward, uncertainty, mentalizing, and learning. The results of our analyses are graphically summarized in Table D.2.

Table D.2 Summary of Neurosynth analysis

Mental process	# of studies	Coronal cut (Dorsal, Ventral)	Sagittal cut (Anterior, Posterior)	Horizontal cut (Left, Right)
Reward	671			
Uncertainty	98			

(continued)

Table D.2 (continued)

Mental process	# of studies	Coronal cut (Dorsal, Ventral)	Sagittal cut (Anterior, Posterior)	Horizontal cut (Left, Right)
Mentalizing	124			
Learning	876			

Notes Search terms: "reward," "uncertainty," "mentalizing," and "learning." The search was conducted on July 21, 2015. "Number of studies" indicates the number of fMRI investigations underlying the results. The brain activation findings illustrated in the pictures (see red accentuation) are based on Neurosynth's default program settings. We recommend that the reader go to the Neurosynth website and use our search terms. Once the results are shown for each mental process, it is possible to use a slide control to show the results for different x, y, and z coordinates

References

Abu-Akel, A. (2003). The neurochemical hypothesis of 'Theory of Mind'. *Medical Hypotheses, 60* (3), 382–386.

Adolphs, R. (2009, September). The social brain: Neural basis of social knowledge, *Annual Review of Psychology, 60,* 693–716.

Adolphs, R., Gosselin, F., Buchanan, T. W., Tranel, D., Schyns, P., & Damasio, A. R. (2005, January). A mechanism for impaired fear recognition after amygdala damage. *Nature, 433* (7021), 68–72.

Adolphs, R., Tranel, D., and Damasio, A. R. (1998, January). The human amygdala in social judgment. *Nature, 393*(6684), 470–474.

Ajzen, I. (1991, December). The theory of planned behavior. *Organizational Behavior and Human Decision Processes, 50*(2), 179–211.

Amodio, D. M., & Frith, C. D. (2006, April). Meeting of minds: The medial frontal cortex and social cognition. *Nature Reviews Neuroscience, 7*(4), 268–277.

Baumgartner, T., Heinrichs, M., Vonlanthen, A., Fischbacher, U., & Fehr, E. (2008, May). Oxytocin shapes the neural circuitry of trust and trust adaptation in humans. *Neuron, 58*(4), 639–650.

Bechara, A., & Damasio, A. R. (2005, May). The somatic marker hypothesis: A neural theory of economic decision. *Games and Economic Behavior, 52*(2), 336–372.

Behrens, T. E. J., Woolrich, M. W., Walton, M. E., & Rushworth, M. F. S. (2007, September). Learning the value of information in an uncertain world. *Nature Neuroscience, 10*(9), 1214–1221.

Behrens, T. E. J., Hunt, L. T., & Rushworth, M. F. S. (2009, May). The computation of social behavior," *Science, 324*(5931), 1160–1164.

Chase, H. W., Kumar, P., Eickhoff, S. B., & Dombrovski, A. Y. (2015). Reinforcement learning models and their neural correlates: An activation likelihood estimation meta-analysis. *Cognitive Affective & Behavioral Neuroscience, 15*(2), 435–459.

Clark, L., Bechara, A., Damasio, H., Aitken, M. R. F., Sahakian, B. J., & Robbins, T. W. (2008, May). Differential effects of insular and ventromedial prefrontal cortex lesions on risky decision-making. *Brain, 13*(5), 1311–1322.

Coleman, J. S. (1990). *Foundations of Social Theory.* Cambridge, MA: Harvard University Press.

Coricelli, G., & Nagel, R. (2009, June) Neural correlates of depth of strategic reasoning in medial prefrontal cortex. In *Proceedings of the National Academy of Sciences (PNAS), 106*(23), 9163–9168.

Critchley, H. D., Mathias, C. J., & Dolan, R. J. (2001, February). Neural activity in the human brain relating to uncertainty and arousal during anticipation. *Neuron, 29*(2), 537–545.

Damasio, A. (2005, April). Human behaviour—Brain trust. *Nature, 435*(7042), 571–572.

Daw, N. D., & Doya, K. (2006, April). The computational neurobiology of learning and reward. *Current Opinion in Neurobiology, 16*(2), 199–204.

Dayan, P., Daw, N. D., & Niv, Y. (2009). Learning, action, inference and neuromodulation. *Encyclopedia of Neuroscience,* 455–462.

Delgado, M. R., Frank, R. H., & Phelps, E. A. (2005, November). Perceptions of moral character modulate the neural systems of reward during the trust game. *Nature Neuroscience, 8*(11), 1611–1618.

Delgado, M. R., & Rigney, A. E. (2009). Reward systems: Human. *Encyclopedia of Neuroscience,* 345–352.

Denny, B. T., Kober, H., Wager, T. D., & Ochsner, K. N. (2012). A meta-analysis of functional neuroimaging studies of self- and other judgments reveals a spatial gradient for mentalizing in medial prefrontal cortex. *Journal of Cognitive Neuroscience, 24*(8), 1742–1752.

Dimoka, A. (2010, June). What does the brain tell us about trust and distrust? Evidence from a functional neuroimaging study. *MIS Quarterly, 34*(2), 373–396.

Doya, K. (2008, April). Modulators of decision making. *Nature Neuroscience, 11*(4), 410–416.

Doya, K., & Kimura, M. (2009). The Basal Ganglia and the Encoding of Value. In P. W. Glimcher, C. F., Camerer, E., Fehr, R. A., & Poldrack (Eds.), *Neuroeconomics: Decision making and the brain* (Vol. 26, pp. 407–416), Amsterdam, Academic Press.

Ellsberg, D. (1961, November). Risk, ambiguity and the Savage axioms. *The Quarterly Journal of Economics, 75*(4), 643–669.

Engell, A. D., Haxby, J. V., & Todorov, A. (2007, September). Implicit trustworthiness decisions: Automatic coding of face properties in the human amygdala. *Journal of Cognitive Neuroscience, 19*(9), 1508–1519.

Fehr, E. (2009a, April–May). On the economics and biology of trust. *Journal of the European Economic Association, 7*(2–3), 235–266.

Fehr, E. (2009b). Social preferences and the brain. In P. W. Glimcher, C. F. Camerer, E. Fehr, & R. A. Poldrack (Eds.), *Neuroeconomics: Decision making and the brain* (pp. 215–232), Academic Press, Amsterdam.

Fehr, E., & Camerer, C. F. (2007, October). Social neuroeconomics: The neural circuitry of social preferences. *Trends in Cognitive Sciences, 11*(10), 419–427.

Fiorillo, C. D., Tobler, P. N., & Schultz W. (2003, March). Discrete coding of reward probability and uncertainty by dopamine neurons. *Science, 299*(5614), 1898–902.

Fishbein, M., & Ajzen, I. (1975). *Belief, attitude, intention, and behavior: An introduction to theory and research.* Reading, MA: Addison-Wesley.

Fletcher, P. C., Happè, F., Frith, U., Baker, S. C., Dolan, R. J., Frackowiak, R. S., et al. (1995, November). Other minds in the brain: A functional imaging study of "theory of mind" in story comprehension. *Cognition, 57*(2), 109–128.

Frisch, D., & Baron, J. (1988, July/September). Ambiguity and rationality. *Journal of Behavioral Decision Making, 1*(3), 149–157.

Frith, C. D., & Frith, U. (2006, May). The neural basis of mentalizing. *Neuron, 50*(4), 531–534.

Frith, U., & Frith, C. D. (2010, January). The social brain: Allowing humans to boldly go where no other species has been. *Philosophical Transactions of the Royal Society–B Biological Sciences, 365*(1537), 165–176.

Frith, U., & Frith, C. D. (2003, March). Development and neurophysiology of mentalizing. *Philosophical Transactions of the Royal Society–B Biological Sciences, 358*(1431), 459–473.

Gallagher, H. L., & Frith, C. D. (2003, February). Functional imaging of 'theory of mind'. *Trends in Cognitive Sciences, 7*(2), 77–83.

Gallagher, H. L., Jack, A. I., Roepstorff, A., & Frith, C. D. (2002, July). Imaging the intentional stance in a competitive game. *NeuroImage, 16*(3), 814–821.

Gallese, V. (2007, April). Before and below 'theory of mind': Embodied simulation and the neural correlates of social cognition. *Philosophical Transactions of the Royal Society-B Biological Sciences, 362*(1480), 659–669.

Garrison, J., Erdeniz, B., & Done, J. (2013). Prediction error in reinforcement learning: A meta-analysis of neuroimaging studies. *Neuroscience and Biobehavioral Reviews, 37*(7), 1297–1310.

Gazzaniga, M. S., Ivry, R., & Mangun, G. R. (2009). *Cognitive Neuroscience: The Biology of the Mind* (3rd ed.). New York: W.W. Norton.

Gefen, D. (2002, August). Nurturing clients' trust to encourage engagement success during the customization of ERP systems. *Omega—International Journal of Management Science, 30*(4), 287–299.

Gefen, D., Benbasat, I., & Pavlou, P. A. (2008, Spring). A research agenda for trust in online environments. *Journal of Management Information Systems, 24*(4), 275–286.

Glimcher, P. W., & Lau, B. (2005). Rethinking the thalamus. *Nature Neuroscience, 8*(8), 983–984.

Gottfried, J. A., O'Doherty, J., & Dolan, R. J. (2003, August). Encoding predictive reward value in human amygdala and orbitofrontal cortex. *Science, 301*(5636), 1104–1107.

Haier, R. J., Siegel, B. V. Jr., MacLachlan, A., Soderling, E., Lottenberg, S., & Buchsbaum, M. S. (1992, January). Regional glucose metabolic changes after learning a complex visuospatial/motor task: A positron emission tomographic study. *Brain Research, 570* (1–2), 134–143.

Hsu, M., Bhatt, M., Adolphs, R., Tranel, D., & Camerer, C. F. (2005, December). Neural systems responding to degrees of uncertainty in human decision-making. *Science, 310*(5754), 1680–1683.

O'Doherty, J., Winston, J., Critchley, H., Perrett, D., Burt, D. M., & Dolan, R. J. (2003). Beauty in a smile: The role of medial orbitofrontal cortex in facial attractiveness. *Neuropsychologia, 41* (2), 147–155.

Jarvenpaa, S. L., Shaw, T. R., & Staples, D. S. (2004, September). Toward contextualized theories of trust: The role of trust in global virtual teams. *Information Systems Research, 15*(3), 250–267.

Javor, A., Ransmayr, G., Struhal, W., & Riedl, R. (2016). Parkinson patients' initial trust in avatars: Theory and evidence. *PLoS ONE, 11*(11), 0165998.

Johnson, T. J., & Kaye, B. K. (2009, January). In blog we trust? Deciphering credibility of components of the Internet among politically interested Internet users. *Computers in Human Behavior, 25* (1), 175–182.

Kenning, P., & Plassmann, H. (2005, November). NeuroEconomics: An overview from an economic perspective. *Brain Research Bulletin, 67*(5), 343–354.

King-Casas, B., Tomlin, D., Anen, C., Camerer, C. F., Quartz, S. R., & Montague, P. R. (2005, April). Getting to know you: Reputation and trust in a two-person economic exchange. *Science, 308*(5718), 78–83.

Knight, F. H. (1921). *Risk, Uncertainty, and Profit* (1st ed.). Hart, Schaffner & Marx, Boston, MA, Houghton Mifflin Company, The Riverside Press, Cambridge.

Knutson, B., Delgado, M. R., & Phillips, P. E. M. (2009). Representation of subjective value in the striatum. In P. W., Glimcher, C. F., Camerer, E., Fehr, & R. A. Poldrack (Eds.), *Neuroeconomics: Decision making and the brain*. Amsterdam: Academic Press.

Kolb, B., & Wishaw, I. Q. (2009). *Fundamentals of Human Neuropsychology* (5th ed.). London: Palgrave Macmillan.

Komura, Y., Tamura, R., Uwano, T., Nishijo, H., Kaga, K., & Ono, T. (2001, August). Retrospective and prospective coding for predicted reward in the sensory thalamus. *Nature, 412*(6846), 546–549.

Krain, A. L., Wilson, A. M., Arbuckle, R., Castellanos, F. X., & Milham, M. P. (2006, August). Distinct neural mechanisms of risk and ambiguity: A meta-analysis of decision-making. *Neuroimage, 32*(1), 477–484.

Krueger, F., McCabe, K., Moll, J., Kriegeskorte, N., Zahn, R., Strenziok, M., et al. (2007, December). Neural correlates of trust. In *Proceedings of the National Academy of Sciences, 104*(50), 20084–20089.

Kuhnen, C. M., & Knutson, B. (2005, September). The neural basis of financial risk taking. *Neuron, 47*(5), 763–770.

Levallois, C., Clithero, J. A., Wouters, P., Smidts, A., & Huettel, S. A. (2012). Translating upwards: Linking the neural and social sciences via neuroeconomics. *Nature Reviews Neuroscience, 13*, 789–797.

Mayer, R. C., Davis, J. H., & Schoorman, F. D. (1995, July). An integrative model of organizational trust. *The Academy of Management Review, 20*(3), 709–734.

Mohr, P. N. C., Biele, G., & Heekeren, H. R. (2010, May). Neural processing of risk. *Journal of Neuroscience. 30*(19), 6613–6619.

Nicolaou, A. I., & McKnight, D. H. (2006, December). Perceived information quality in data exchanges: Effects on risk, trust, and intention to use. *Information Systems Research, 17*(4), 332–351.

Niv, Y., Montague, P. R. (2009). Theoretical and empirical studies of learning. In P. W. Glimcher, E., Fehr, A., Rangel, C., Camerer, & A. P., Russell (Eds.), *Neuroeconomics: Decision making and the brain*. Amsterdam: Academic Press.

O'Doherty, J. P. (2004, December). Reward representations and reward-related learning in the human brain: Insights from neuroimaging. *Current Opinion in Neurobiology, 14*(6), 769–776.

O'Doherty, J. P. (2009). Reward processing: Human imaging. *Encyclopedia of Neuroscience*, 335–343.

O'Doherty, J., Dayan, P., Schultz, J., Deichmann, R., Friston, K., & Dolan, R. J. (2004, April). Dissociable roles of ventral and dorsal striatum in instrumental conditioning. *Science, 304* (5669), 452–454.

Pavlou, P. A., & Gefen, D. (2004, March). Building effective online marketplaces with institution-based trust. *Information Systems Research, 15*(1), 37–59.

Pavlou, P. A., Liang, H., & Xue, Y. (2007, March). Understanding and mitigating uncertainty in online exchange relationships: A principal-agent perspective. *MIS Quarterly, 31*(1), 105–136.

Perner, J., & Lang, B. (1999, September). Development of theory of mind and executive control. *Trends in Cognitive Sciences, 3*(9), 337–344.

Phillips, M. L., Young, A. W., Scott, S. K., Calder, A. J., Andrew, C., Giampietro, V., et al. (1998, October). Neural responses to facial and vocal expressions of fear and disgust. In *Proceedings of the Royal Society of London—B Biological Sciences, 265*(1408), 1809–1817.

Phillips, M. L., Young, A. W., Senior, C., Brammer, M., Andrew, C., Calder, A. J., et al. (1997, October). A specific neural substrate for perceiving facial expressions of disgust. *Nature, 389* (6650), 495–498.

Plassmann, H., Kenning, P., Deppe, M., Kugel, H., & Schwindt, W. (2008, July–October). How choice ambiguity modulates activity in brain areas representing brand preference: Evidence from consumer neuroscience. *Journal of Consumer Behaviour, 7*(4–5), 360–367.

Platt, M. L., & Huettel, S. A. (2008, April). Risky business: The neuroeconomics of decision making under uncertainty. *Nature Neuroscience, 11*(4), 398–403.

Ploghaus, A., Tracey, I., Gati, J. S., Clare, S., Menon, R. S., Matthews, P. M., et al. (1999, June). Dissociating pain from its anticipation in the human brain. *Science, 284*(5422), 1979–1981.

Premack, D., & Woodruff, G. (1978). Does the chimpanzee have a theory of mind? *Behavioral and Brain Sciences, 1*(04), 515–526.

Preuschoff, K., Bossaerts, P., & Quartz, S. R. (2006, August). Neural differentiation of expected reward and risk in human subcortical structures. *Neuron, 51*(3), 381–390.

Preuschoff, K., Quartz, S. R., and Bossaerts, P. (2008, March). Human insula activation reflects risk prediction errors as well as risk. *Journal of Neuroscience, 28*(11), 2745–2752.

Price, C. J., & Friston, K. J. (2005, June). Functional ontologies for cognition: The systematic definition of structure and function. *Cognitive Psychology, 22*(3–4), 262–275.

Rescorla, R. A., & Wagner, A. R. (1972). A theory of Pavlovian conditioning: Variations in the effectiveness of reinforcement and nonreinforcement. In A. H. Black & W. F. Prokasy (Eds.), *Classical conditioning II: Current research and theory* (pp. 64–99), New York: Appleton-Century-Crofts.

Riedl, R., Banker, R. D., Benbasat, I., Davis, F. D., Dennis, A. R., Dimoka, A. (2010a). On the foundations of NeuroIS: Reflections on the Gmunden Retreat 2009. *Communications of the Association for Information Systems, 27*(15), 243–264.

Riedl, R., & Javor, A. (2012). The biology of trust: Integrating evidence from genetics, endocrinology, and functional brain imaging. *Journal of Neuroscience, Psychology, and Economics, 5*(2), 63–91.

Riedl, R., Hubert, M., & Kenning, P. (2010b, June). Are there neural gender differences in online trust? An fMRI study on the perceived trustworthiness of eBay offers. *MIS Quarterly, 34*(2), 397–428.

Riedl, R., Mohr, P., Kenning, P., Davis, F. D., & Heekeren, H. (2011). Trusting humans and avatars: Behavioral and neural evidence. In *Proceedings of the 32nd International Conference on Information Systems (ICIS)*, Shanghai, China, December 4–7.

Riedl, R., Mohr, P., Kenning, P., Davis, F., & Heekeren, H. (2014). Trusting humans and avatars: A brain imaging study based on evolution theory. *Journal of Management Information Systems, 30*(4), 83–113.

Riegelsberger, J., Sasse, M. A., & McCarthy, J. D. (2005, March). The mechanics of trust: A framework for research and design. *International Journal of Human-Computer Studies, 62*(3), 381–422.

Rushworth, M. F. S., Behrens, T. E. J., & Walton, M. E. (2007). The anterior cingulate cortex in learning and reward-guided decision making. *Neuroscience Research, 58*(1), S22.

Saxe, R., & Kanwisher, N. (2003). People thinking about thinking people: The role of the temporo-parietal junction in 'theory of mind'. *NeuroImage, 19*(4), 1835–1842.

Schultz, W. (2002, October). Getting formal with dopamine and reward. *Neuron, 36*(2), 241–263.

Schultz, W. (2000, December). Multiple reward signals in the brain. *Nature Reviews Neuroscience, 1*(3), 199–207.

Schultz, W. (2009). Midbrain Dopamine Neurons: A Retina of the reward system? In P. W., Glimcher, E., Fehr, A., Rangel, C., Camerer, & R. A., Poldrack (Eds.), *Neuroeconomics: Decision making and the brain* (pp. 321–327). Amsterdam: Academic Press.

Schultz, W. (2006, January). Behavioral theories and the neurophysiology of reward. *Annual Review of Psychology, 57*, 87–115.

Schultz, W., Preuschoff, K., Camerer, C. F., Hsu, M., Fiorillo, C. D., Tobler, P. N., & Bossaerts, P. (2008, December). Explicit neural signals reflecting reward uncertainty. *Philosophical Transactions of the Royal Society B-Biological Sciences, 363*(1511), 3801–3811.

Schultz, W., Dayan, P., & Montague, P. R. (1997, March). A neural substrate of prediction and reward. *Science, 275*(5306), 1593–1599.

Shimojo, S., Simion, C., Shimojo, E., and Scheier, C. (2003, November). Gaze bias both reflects and influences preference. *Nature Neuroscience, 6*, 1317–1322.

Sidorova, A., Evangelopoulos, N., Valacich, J. S., Ramakrishnan, T. (2008, September). Uncovering the intellectual core of the information systems discipline. *MIS Quarterly, 32*(3), 467–482.

Simmons, A., Matthews, S. C., Stein, M. B., & Paulus, M. P. (2004). Anticipation of emotionally aversive visual stimuli activates right insula. *Neuroreport, 15*(14), 2261–2265.

Singer, T. (2009). Understanding others: Brain mechanisms of theory of mind and empathy. P. W., Glimcher, E., Fehr, A., Rangel, C., Camerer, & R. A., Poldrack (Eds.), *Neuroeconomics: Decision making and the brain* (pp. 251–268). Academic Press, Amsterdam.

Singer, T., Critchley, H. D., & Preuschoff, K. (2009). A common role of insula in feelings, empathy and uncertainty. *Trends in Cognitive Sciences, 13*(8), 334–340.

Smith, D. V., and Delgado, M. R. 2015. "Reward processing," *Brain Mapping: An Encyclopedic Reference*, A. W. Toga (ed.), Academic Press, Waltham, MA, pp. 361–366.

Steininger, K., Riedl, R., Roithmayr, F., & Mertens, P. (2009). Fads and trends in business and information systems engineering and information systems research: A comparative literature analysis. *Business & Information Systems Engineering, 1*(6), 411–428.

Teo, T. S. H., Srivastava, S. C., & Jiang, L. (2008, Winter). Trust and electronic government success: An empirical study. *Journal of Management Information Systems, 25*(3), 99–131.

Todorov, A. (2008). Evaluating faces on trustworthiness: An extension of systems for recognition of emotions signaling approach/avoidance behaviors. *Annals of the New York Academy of Sciences, 1124*, 208–224. (The Year in Cognitive Neuroscience 2008).

Todorov, A., Baron, S. G., & Oosterhof, N. N. (2008, March). Evaluating face trustworthiness: A model based approach. *Social Cognitive and Affective Neuroscience, 3*(2), 119–127.

Tricomi, E. M., Delgado, M. R., & Fiez, J. A. (2004, January). Modulation of caudate activity by action contingenc. *Neuron, 41*(2), 281–292.

Van Overwalle, F. (2009, March). Social cognition and the Brain: A meta-analysis. *Human Brain Mapping, 30*(3), 829–858.

Vessey, I., Ramesh, V., & Glass, R. L. (2002, Fall). Research in information systems: An empirical study of diversity in the discipline and its journals. *Journal of Management Information Systems, 19*(2), 129–174.

Wang, W. Q., & Benbasat, I. (2008). Attributions of trust in decision support technologies: A study of recommendation agents for e-commerce. *Journal of Management Information Systems, 24*(4), 249–273.

Wheeler, E. Z., & Fellows, L. K. (2008, March). The human ventromedial frontal lobe is critical for learning from negative feedback. *Brain, 131*(5), 1323–1331.

Willis, J., & Todorov, A. (2006, July). First impressions: Making up your mind after 100 ms exposure to a face. *Psychological Science, 17*(7), 592–598.

Wimmer, H., & Perner, J. (1983, January). Beliefs about beliefs: Representation and constraining function of wrong beliefs in young children's understanding of deception. *Cognition, 13*(1), pp. 103–128.

Winston, J. S., Strange, B. A., O'Doherty, J., and Dolan, R. J. (2002, March). Automatic and intentional brain responses during evaluation of trustworthiness of faces. *Nature Neuroscience, 5*(3), 277–283.

Yacubian, J., Gläscher, J., Schroeder, K., Sommer, T., Braus, D. F., & Büchel, C. (2006, September). Dissociable systems for gain- and loss-related value predictions and errors of prediction in the human brain. *Journal of Neuroscience*, 26(37), 9530–9537.

Zak, P. J. (2003, August). Trust. *Capco Institute Journal of Financial Transformation, 7*, 17–23.

Printed in the United States
By Bookmasters